T0268861

A BRIEF HISTORY OF STUFF

The Extraordinary Stories of Ordinary Objects

DK LONDON
Editor Millie Acers
Designer Joelle Wheelwright
Senior Art Editor Anna Formanek
Senior Acquisitions Editor Pete Jorgensen
Managing Art Editor Jo Connor
Production Editor Marc Staples
Production Controller Louise Minihane
Managing Director Mark Searle

DK would like to thank Madeline Pollard for copyediting, Caroline Curtis and Jen Moore for proofreading and Melanie Gee for indexing.

Special thanks to Tim Leng, Wendy Burford and Jenny Lawson for their expertise and editorial excellence.

First American Edition, 2024
Published in the United States by DK Publishing,
a division of Penguin Random House LLC
1745 Broadway, 20th Floor, New York, NY 10019

24 25 26 27 28 10 9 8 7 6 5 4 3 2 1
001–333562–Jun/2024

Published in Great Britain by Dorling Kindersley Limited

A catalog record for this book
is available from the Library of Congress.
ISBN 978-0-7440-8160-2

Printed and bound in China

www.dk.com

Foreword by

Nihal Arthanayake

CONTENTS

MENU

FOREWORD

The wonders of science are so often promoted by headline-grabbing stories of human ingenuity which are either grand in scale, so that we can marvel at each technological spectacle in all its glory, or which give us a tantalising glimpse of the future by turning science fiction into tangible plausibility. We are drawn to these stories because each one celebrates great moments in the onward march of human progress and provides us with positive narratives to cling onto. But while we wait for each new scientific, industrial and technological development to be heralded on social media and our news apps, we might well be ignoring the fascinating scientific stories that lie behind what is disparagingly referred to as the "stuff" that surrounds us.

As you walk through your front door in the evening and switch on the light, what do you know of the history of the lightbulb that illuminates your living spaces? If you cannot be bothered to cook, and reheating last night's takeaway in the microwave appears to be the only option, what thought do you give to the story of that ubiquitous domestic appliance as the aromatic curry spins slowly around and you hungrily wait for the ping that tells you dinner is ready? How aware are you of the captivating account of its design and construction?

From the story of the stockings that cling to our legs and the curling tongs that twist and shape our hair into a plethora of different styles, to the matches you strike when lighting a candle and the mobile phones that have become an accessory we can barely do without, it is time for you to delve into *A Brief History of Stuff*. As you read on you will quickly realise how much these innocuous objects that we take for granted have helped alter the path of human development. When Europeans invaded the Americas in the 15th century and noticed the Indigenous people playing with a ball made of a tough, waterproof substance that they had never encountered before, nobody could have predicted how vital the substance that came from "weeping trees" would be in providing humans with protection from STIs and HIV centuries later. Is anyone aware of the connection between the roller skate and the American Civil Rights movement? As part of the global attempt to fight the climate crisis, countries across the planet will have to replace petrol and diesel vehicles with electric ones. You may be surprised to learn that you could have hailed an electric taxi on the streets of London at the end of the 19th century.

In *A Brief History of Stuff* you will witness how science gleans knowledge from the unlikeliest of sources, a perfect example of which is the fact that the Science Museum has a rubber duck within its vast collection of objects. Why? Because this simple bath toy tells us something about ocean surface currents. The ordinary becomes extraordinary when you are exposed to the ideas and innovations that have taken place over millennia, whether they are to help with menstrual periods, or provide visibility for those whose eyes don't work as well as they once did.

They also remind us of the power of collaboration, whether that be contemporaries working together in a laboratory, or knowledge handed to us from way back in antiquity that forms the very foundation of what you hold in your hand today. As someone getting ready to go out waits for their hair straighteners to warm up, do they know that over three and a half thousand years ago the Egyptians used bronze to make an implement that both curled and trimmed hair? This is just one of the many weird and wonderful stories of the objects that surround us.

By the time you finish *A Brief History of Stuff* you will revel in the wonder contained within the seemingly mundane and find a new respect for the ruler in a pencil case, the first aid kit in your cupboard, or the trusty office chair. Just because an object isn't bold, brash and big, that doesn't mean there isn't a captivating story behind its inception, evolution and cultural significance. Read on, and you will discover exactly what I mean.

Nihal Arthanayake,
Host of the *Brief History of Stuff* podcast

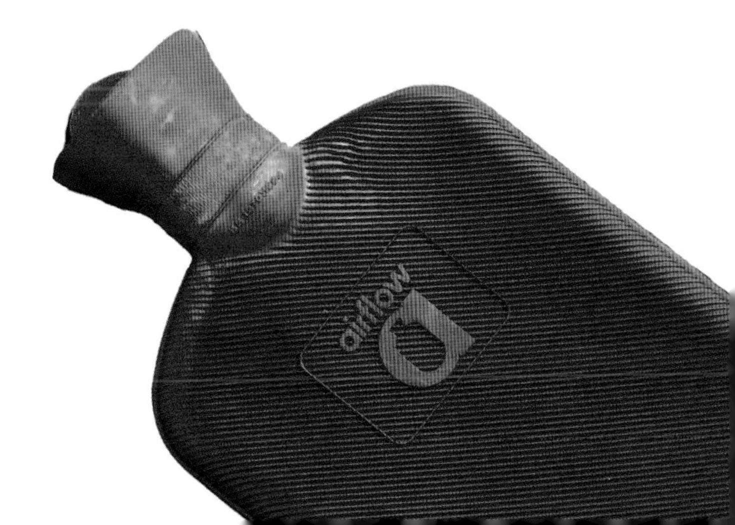

ELECTRIC TAXI

Electric cars are nothing new.

As more and more are seen out on the roads, and are celebrated by the entrepreneurs and tech bros of Silicon Valley as being the way forward, it's hard to imagine that these electric vehicles have actually been around for centuries.

If you were part of the wealthy elite that could afford to buy a car in the 1890s, you had three options: steam, petrol or electric. Each of them had their drawbacks. Steam cars required enormous amounts of water to run, and needed half an hour for the engine to heat up before you could go anywhere. Petrol cars at this time were blighted by the need to hand-crank, which involved turning a large lever at the front of the car to start the engine – a process that could easily break your wrist if the engine did not engage properly. However, there was an alternative: the electric car.

In late 19th-century New York, 90 per cent of taxis were electric, as were more than 40 per cent of cars on American roads. The market seemed set for the electric car and London saw its first electric taxi in 1897.

The taxi in question was the Bersey Electric taxi, designed by Walter Bersey. An engineer by trade, Bersey spent the early part of his career developing electric vehicles. His taxi could hold two passengers and had a heady top speed of 11 mph (about 17 km/h). This was slower than the petrol cars of the day, but still ideally suited to the pace of London's crowded streets. In praise of his invention, Bersey remarked, "there is no smell, no noise, no heat, no vibration, no possible danger, and it has been found that vehicles built on this company's system do not frighten passing horses." (The latter was an important concern in London at the time, as horse and carriages were still a common way to get around.) Before it was given a licence by Scotland Yard, it was put to the test on the steepest hill in London, the Savoy Hill on the Strand, passing the test with flying colours. Its yellow and black livery and the hum of its batteries meant it was nicknamed "the hummingbird".

"In late 19th-century New York, 90 per cent of taxis were electric, as were more than 40 per cent of cars on American roads."

The Bersey's battery was made up of 40 "dry cell" batteries, weighing three-quarters of a tonne. Dry cell

In total, around 77 Bersey Electric taxis were made, but the example in Science Museum Group's collection is thought to be one of only a few to survive.

batteries enabled chemical energy to be converted into electricity: an electrolyte (such as carbon or manganese dioxide) reacts with a substance (for example, zinc) to create electricity. Electricity leaves the battery via positive and negative electrodes, and is then used to power the vehicle.

Its quiet, easy-to-start engine appeared to give the Bersey an advantage over its petrol and steam-propelled rivals. However, its time on the streets of London was to be short-lived. Its range of just 35 miles (56 km) was barely enough to complete a day's work for a London taxi driver, who needed vehicles with longer ranges if they were to surpass horses. The Bersey also required a hydraulic lifting system to change its batteries, and there was just one recharging system in the whole of London. Furthermore, the taxi would begin to vibrate violently, wearing through tyres and other key components in just a couple of months. It soon became clear that this was a far from ideal way to transport passengers. The last Bersey was withdrawn from operation in 1899.

The invention of improved starting mechanisms meant that petrol cars quickly took control of the market. Electricity at the time was expensive, and, because of their limited range and the scarcity of refuelling stations, it was impossible to drive electric cars in rural settings. Petrol cars had much longer ranges and early petrol stations could consist of just a large tank and a hand pump, compared to the complex infrastructure needed at electric car charging stations. An early electric car took hours to charge, as opposed to the minutes taken to fuel a petrol

The 1897 Bersey Electric taxi and its driver.

car. While electric-powered vehicles did not completely disappear (electric milk floats were common from 1932 onwards), petrol and diesel cars came to dominate the market. In 1950 there were just 4 million vehicles on the road, but as disposable incomes increased and car prices dropped this figure grew to 34 million by 2010. This in turn meant more reliance on fossil fuels as well as their polluting effects. However, as passengers and drivers both seek more environmentally friendly and cheaper forms of transport, electric cars have increased in popularity.

Climate change has forced the world to phase out fossil fuels, and electric vehicles have emerged as a sustainable form of transport. In Norway, a country where 95 per cent of electricity produced is "green electricity", more than 75 per cent of cars sold in 2020 were electric. And after 120 years, the electric taxi returned to London's streets in 2019 with the Nissan Dynamo. In the UK in 2022, electric vehicles accounted for 16 per cent of all new car sales.

The 2019 Nissan Dynamo, London's all-electric black cab.

While electric cars are now presented as the shiny new thing of the present, it is interesting to consider how steam, electric and petrol once shared the same streets. Electric cars have gone from being second best to petrol and diesel, to reaching the mainstream. As we look back on their history, it is worth asking the question: where would we be now if more efficient batteries had been invented in the early 20th century?

CONDOMS

In the 15th century, European invaders observed the Indigenous people of the Americas playing with a ball that possessed extraordinary qualities: it was incredibly bouncy, never seemed to get wet and rarely broke. Civilisations from across the Caribbean and North and South America – including the Aztecs, the Olmecs and the Maya of Mesoamerica – had discovered a milky fluid from local trees. When combined with juice from morning glory plants, this milky white liquid, which we now know as latex, could produce a waterproof and highly resilient material: rubber.

According to Spanish conquistadors, this material was used in a variety of contexts, from bouncy balls to rain shoes. Today, rubber, and specifically rubber made from latex, continues to be used in many aspects of our day-to-day lives, including during some of our most intimate moments. The condom has become one of the most common contraceptives in the world, also offering a "barrier" protection method against sexually transmitted infections (STIs), with more than 160 million sold each year in the UK alone – that's 3 million a day.

Worries about STIs, and the desire to have greater control over when to reproduce, have existed throughout human history. Barrier protection – referring to protective coverings placed over the penis or receptacles placed inside the vagina – is one of the oldest ways of safeguarding against STIs or unwanted pregnancies. For many years, the anatomist Gabriele Falloppio (who gives his name to the fallopian tube) was credited with inventing the first condom in Europe in the 1550s or 1560s. However, archaeological evidence shows that humans have been using barrier contraception for thousands of years.

"Worries about STIs, and the desire to have greater control over when to reproduce, have existed throughout human history."

The first known record of a protective "sheath" being used during sexual intercourse is from the Bronze Age, around 5,000 years ago, in relation to King Minos of Crete. However, the technology of these "sheaths" was nowhere near the quality of the latex-based condoms of today. The ancient Egyptians and Romans used linen or cloth that was often soaked or filled with herbs, poultices and other medicines believed to stop disease and function as contraception (sometimes surprisingly effectively).

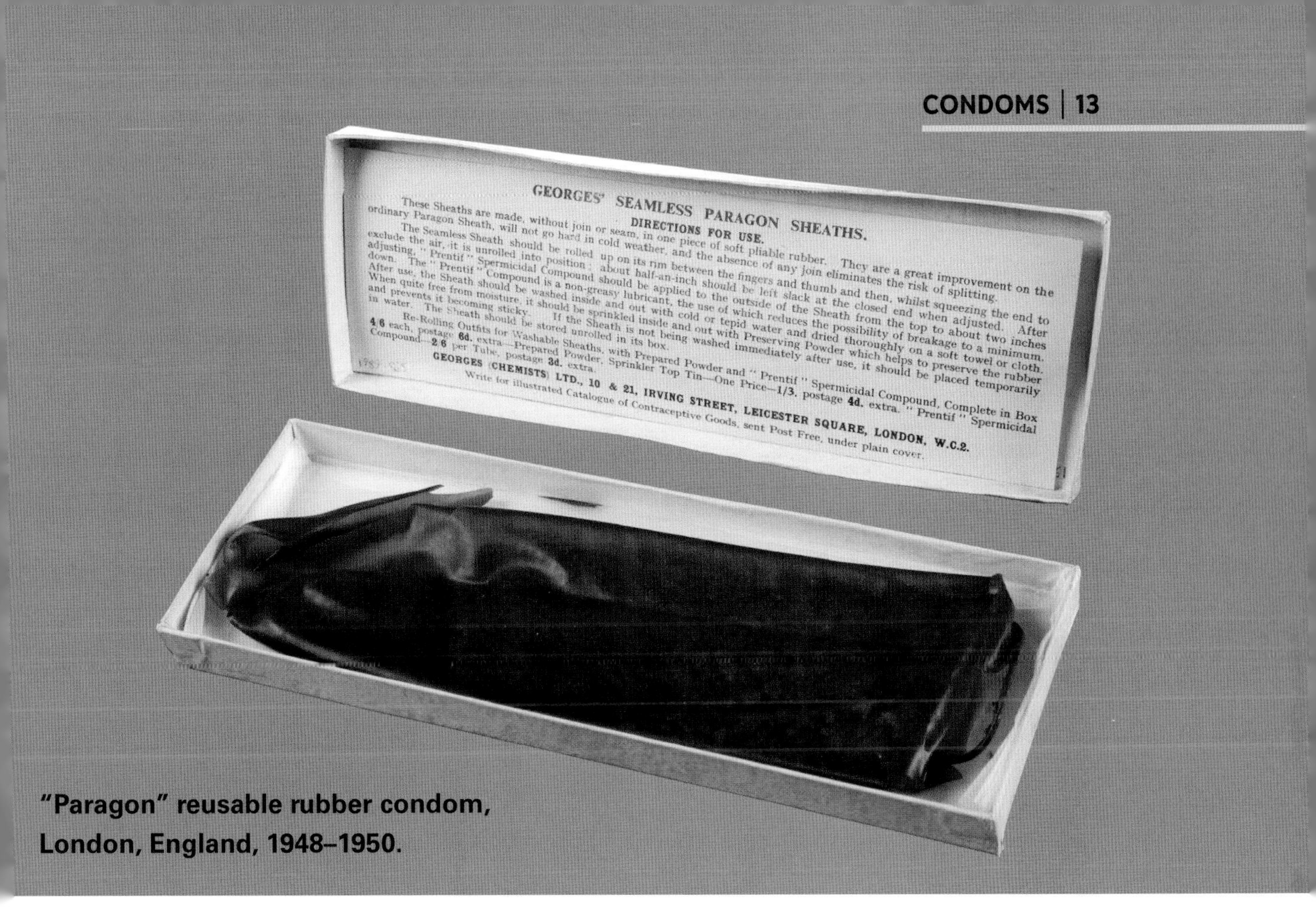

"Paragon" reusable rubber condom, London, England, 1948–1950.

Chinese cultures before the 15th century used similar technology, but instead of cloth or linen opted for oiled or silk paper. In Japan, it was common for sheaths to be made of leather or even tortoise shell. Animal organs were also used, particularly intestines or bladders, as they stretched more comfortably and transferred heat for a more pleasurable experience. However, many of these examples – all designed to be washed and reused – would have been expensive to buy and may have been sewn with a seam that risked leaking or tearing.

For centuries, condom technology remained largely the same, even when King Charles II asked for one to be specially made for him. While natural rubber showed promise, it could become sticky or brittle, and even rotted over time, which meant using it for anything delicate was near impossible. This changed in 1839 when Charles Goodyear – of what would become the international corporation Goodyear Tires & Rubber Company – made a startling discovery. If rubber was heated and sulphur was added, the finished material had higher tensile strength, resisted damage, and could keep its flexibility and elasticity over broad temperatures. Rubber condoms made it to market by 1855, along with other consumer goods from tyres to hot water bottles. They redefined expectations about the look and accessibility of barrier protection, were more durable (lasting around three months) and easily washable (which was still the common practice then, but not advised today). They could also be produced and sold at a much lower cost.

Seamless condoms were created at the turn of the 20th century. These were a significant improvement on their predecessors: up until this point, rubber condoms still had the issue of being sewn together, and were often very thick, making them bulky and unpleasant to wear or insert. Seamless rubber condoms continued to be used for many decades, including the example from the previous page dating between the 1940s and '50s.

In the 1920s and '30s, rubber latex was invented. Rubber latex, often simply called latex, is produced by combining liquid latex with a variety of chemicals that keep the latex liquid when heated at a certain temperature. This liquid can then be poured over glass moulds, treated, washed, and then gently hardened into an even more elasticated form that lasts for up to five years. The modern condom was born.

Despite evidence that the condom was an effective way to reduce rates of STIs and protect against unplanned pregnancy, condoms were illegal in the USA from 1873 to 1918 due to the Comstock laws. These laws banned "every article or thing designed, adapted, or intended for preventing conception" in order to prevent what was considered to be promoting "immoral purpose". The US Army lost about 7 million days of active duty due to active STIs, a fact

"By 1855, rubber condoms made it to the market, and redefined expectations about the look and accessibility of barrier protection."

The white sap from rubber trees is used to create latex.

that perhaps contributed to the decision to legalise condoms and even give out condom rations to soldiers during World War II. With high demand came innovation, and developments in the design and technology of the condom really took off from the 1930s onward. In the 1940s, Japanese manufacturers would develop the first coloured latex condoms and, in 1957, the British brand Durex would produce the first ever lubricated option.

Up until this time, the primary purposes for using the condom were to prevent unwanted pregnancy and to stop the spread of the (then very deadly) diseases syphilis and gonorrhoea. The 1960s saw penicillin become much more widely available, curbing these bacterial infections. When the contraceptive pill was introduced in 1961, family planning was no longer reliant on barrier contraception. For decades, the use of the latex condom was on a steady decline.

Then came the HIV/AIDS crisis. At first, it was unclear what this illness was and how it was spreading. It was suggested in 1982 that this viral infection could be sexually transmitted and that only barrier protection could effectively protect against it, but governments were reluctant to encourage condom usage, associating condoms with immoral, licentious behaviour. Despite the US Surgeon General Dr C. Everett Koop supporting condom promotion, President Ronald Reagan concentrated on promoting abstinence. Some people even went so far as to suggest that AIDS victims – mostly gay men, sex workers and drug users – were getting what they deserved.

This didn't stop large-scale advertising campaigns promoting condom usage in print media across the US and Europe. Governments, at first slow to support these campaigns, eventually got behind them. Adverts for condoms hit UK television sets in November 1987, with the United States following suit in 1991. Having governments invest in encouraging condom usage was a turning point in messaging around sexual health. The year after the British television campaign began, condom sales in the UK were up by 20 per cent. In 1988, condoms were the most popular birth control among married couples in Britain for the first time following the invention of the contraceptive pill. Popularity in the USA was slightly less; by the 1990s they were the third most popular choice for married couples, and the second most popular choice among single women.

Since this huge increase in demand, condoms have undergone new and spectacular designs, with options ranging from glow-in-the-dark to flavoured and even chicken shaped! Condoms have also been mass-produced from plastics – either polyurethane or polyisoprene – since the 1990s, catering for those who have allergic reactions to latex products. However, the original rubber latex condoms continue to be used more widely, and are slightly less likely to break than their purely plastic counterparts.

When used correctly, modern condoms are 98 per cent effective against pregnancy and the spread of STIs. This is due to the extraordinary technology behind the development of rubber and latex. Condoms have also had a long-lasting impact on society and public health across the globe for the last century. The Joint United Nations Programme on HIV and AIDS estimates that 45 million HIV infections have been prevented by condom use, helping millions of people protect themselves and their sexual partners.

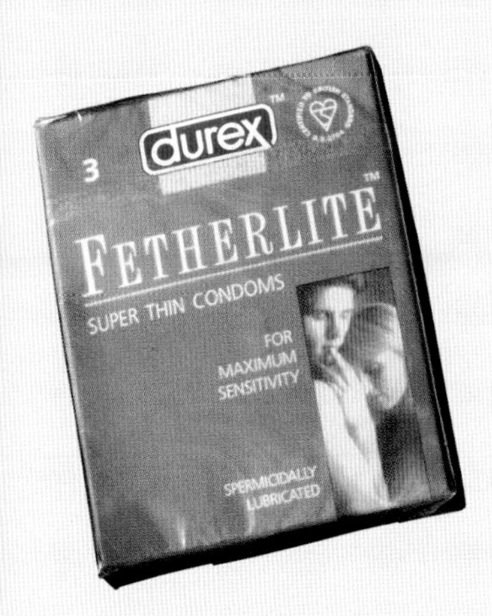

A collection of flavoured and coloured Durex condoms from 1995.

BABY'S BOTTLE

The decision to breastfeed or not is a deeply personal one. There have always been complex reasons why individuals and families may be unable to breastfeed or choose not to. For some, the availability of formula milk can be lifesaving. For others, a combination of breastfeeding and formula feeding works best. From prehistory to the modern day, feeding bottles, like the one pictured opposite, have been a familiar ritual of life for many. While their design, material and appearance may have changed, they all serve the same purpose: trying to give babies the best possible start in life.

How babies are fed – whether with formula, animal or breast milk (or a combination of these) – has varied hugely across time and place. Every culture has their own understanding of how best to care for babies, which is reflected in the popularity of different feeding practices throughout history.

What baby bottles are made of is as important as what is in them, particularly when it comes to keeping them clean. Early feeding vessels were often made of metal or ceramic. Several ceramic feeding vessels have been found in the graves of infants and children dating back to the Stone Age, for example. They were not bottle-shaped as we would recognise it today. For thousands of years, these baby feeders tended to have small openings and tight corners, which were difficult to clean. Ceramic is also a porous material, meaning it can trap bacteria and create a health risk. It was not widely understood that germs and poor hygiene could spread disease until the late 1800s.

In the Victorian era, the "nipples" used were particularly problematic and often also harboured bacteria. They were made of rubber or animal teats and commonly attached to a long rubber tube, which allowed the infant to hold the bottle. Household manuals like *Mrs Beeton's Book of Household Management* (1861) recommended that these nipples could be left on the bottle until they were replaced every couple of weeks, and only lightly rinsed in the meantime. This was one factor that contributed to the high infant mortality rate of the time. In 1841, around 15 per cent of babies in the UK died before their first birthday.

"What baby bottles are made of is as important as what is in them, particularly when it comes to keeping them clean."

Glass infant's feeding bottle with cleaning brush in original box, griptight miniature feeder, c. 1940.

The 1940s "Grip-tight" glass baby bottle pictured on the previous page includes original accessories like a bottle brush and a nipple and stopper for each end of the bottle. It is see-through, unlike earlier ceramic or metal feeders. This made it much easier to monitor its cleanliness. Its boat shape, and the opening at each end, also meant there were fewer nooks and crannies that were difficult to scrub and could harbour germs. The inclusion of a bottle brush is a clear sign that the need for good sanitation was understood. By the time this bottle was developed in the '40s, the nipple and stopper were also made from rubber that was sturdy enough to be sterilised at high heat. These safety improvements contributed to an increase in bottle feeding, which reached a peak in the 1970s before the rise in campaigns to encourage breastfeeding.

Nowadays, baby bottles are generally made of plastic and are designed to be easy to clean and sterilise. A huge range of bottle designs are available, from those shaped to try to reduce colic, to bottles that aim to mimic the size and shape of breasts. However, safe and effective formula feeding today still depends on the availability of sterilising equipment and access to clean water for washing. Discussions continue about the best materials to use, with plastics containing BPA (bisphenol A, a chemical used in the production of polycarbonate plastics) banned in the EU since 2011. Research found that when used for food and drink containers, BPA could seep into the contents. Prolonged exposure to this has potential health impacts.

"While Franz Simon was the first to chemically compare human and animal milk, people had long observed that babies did not thrive as well when fed just animal milk."

The first powdered infant formula was developed in the 1860s. The chemical differences between cow's milk and human breast milk were first identified in the 1830s by German scientist Johann Franz Simon. He found that cow's milk had more protein and fewer carbohydrates than breast milk. The influence of his analysis could be seen in the infant formulas made as late as the 1940s, which was based on evaporated cow's milk and contained added sugars to increase the amount of carbohydrates. Before pre-made infant formula became widely available in the mid-1900s, people added things like orange juice to animal milk to try to enrich it. While Franz Simon was the first to chemically compare human and animal milk, people had long observed that babies did not thrive as well when fed just animal milk.

Since the 1940s, research has continued to try to make formula milk as similar to human breast milk as possible. For example, scientists discovered that there are particular sugar molecules in breast milk that are not digested by the baby, but which encourage the growth of healthy bacteria in the gut. These sugars are less abundant in cow's milk, but they can now be made and added to baby formula. There are, however, important elements of breast milk that can't yet be replicated. When babies are breastfed, antibodies in the milk can help to build their immunity to common diseases.

It is worth remembering that the decision to use formula rather than breastmilk is often politicised. Formula companies have even been accused of marketing their products in unethical ways, exploiting the lack of breastfeeding-friendly policies to sell their product or even spreading misinformation about the superiority of formula milk. In 1977, consumers in the US and Europe launched a boycott against Nestlé in protest at their aggressive marketing of formula milk, particularly in developing countries.

For as long as powders and milks have been mixed with water to make food for infants, access to clean water has been vital for the safety of the practice. Before the introduction of modern water and sanitation systems, this was a problem around the world, and there are still significant issues accessing clean water in many countries today. It is particularly problematic in developing nations, but as recently as 2016, the level of lead in the water in Flint, Michigan, US, made it dangerous to use tap water to make infant formula.

Of course, baby feeding instruments are not only used with formula milk. Designs for breast pumps were first patented more than 100 years ago, but these machines were

Significant research has been done to ensure modern baby formula is as similar to human breast milk as possible.

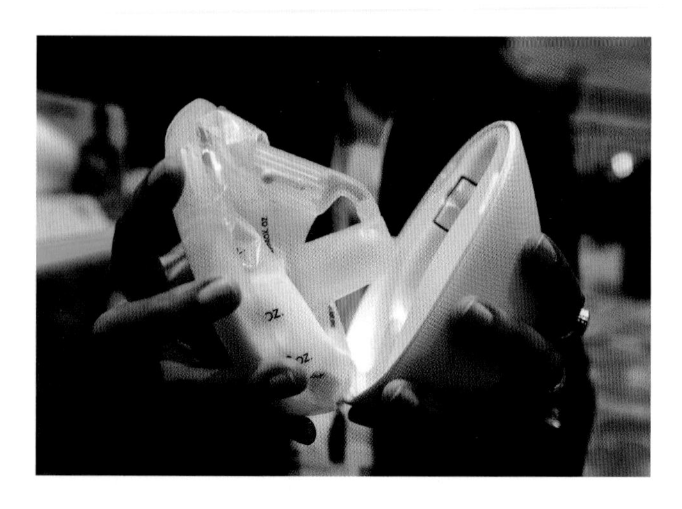

A "smart" breast pump from Willow. Released in 2017, this wearable pump is marketed as quiet enough to use during a conference call.

painful, inefficient and often bulky. It was not until the 1980s and '90s that convenient portable breast pumps became readily available. This made bottle feeding using expressed breast milk much more practical. The development of fridges and freezers also allowed for the safe storage of this milk. Pumping is particularly important in places like the US, where there is no legal requirement for companies to offer maternity leave, and many employees go back to work soon after giving birth.

Formula milk or animal milk have not always been the only alternatives to a parent's own breast milk. For hundreds, if not thousands, of years, wet nursing was commonly practised, although how it was organised and who could access it changed with the times. Between the 1500s and 1800s, European medical texts frequently stated that, among other requirements, wet nurses should be cheerful and virtuous, and have good teeth. In the UK, upper-class families often sent their babies to the countryside for wet nursing. In the 1800s, in the southern states of the US, enslaved Black people were commonly used as wet nurses. This was another way of controlling the lives and bodies of enslaved people, and could have severe consequences for the wet nurse's own children. Enslaved wet nurses were often forced to wean their babies very early, or prioritise feeding their enslavers' children. Some were also forced to act as wet nurses after their own babies had died, with no thought for the emotional impact of this.

Directly wet nursing other people's children is uncommon today, but there are still ways that breast milk is shared. Breast milk donations can have a huge impact in a neonatal intensive care unit (NICU), a nursery in a hospital that provides around-the-clock care to sick or preterm babies, where breastfeeding can be difficult for a range of reasons. Sick babies are sometimes transferred to specialist hospitals a long way from home, for example, and the stress of having an ill child can also make producing breast milk difficult, interrupting supply in the parent. Donated breast milk can make a real difference in these instances, providing all the benefits of breast milk, such as protection from necrotising enterocolitis, a serious gut condition that mainly affects premature babies.

Trends, research and knowledge about baby feeding, and attitudes towards it, have varied enormously throughout history. How to feed an infant is a deeply personal and often very emotional decision. This baby bottle, and other historical examples like it, show that while this technology continues to evolve, the ultimate goal remains the same: healthy and thriving babies.

COSMETICS

"Imparting a delicate and permanent whiteness to the face, neck, arms", promises the label of the glass bottle shown overleaf, a facial cosmetic product from the late 1800s and early 1900s similar to modern foundation. It's part of a complex history of cosmetics shaped by gender, sexuality, class, racism, colonialism, consumer capitalism, chemistry and medicine.

Today, skin whitening, lightening or bleaching is a widespread global phenomenon. In the UK, it is mostly practised by people of African, Caribbean and Asian heritage, while skin treatments such as tanning and bronzing are often sought after by the white population. Seemingly opposite, these practices exist within the systemic racism of a multi-billion-dollar beauty industry. It is well known that the mainstream beauty industry does not cater for dark skin. In 2021, Black people in the US were the fastest-growing segment of beauty consumers, spending $6.6 billion. Nevertheless, they were three times more likely to be dissatisfied with the products available to them than their non-Black counterparts. Though there are Black-owned businesses, which are likely to be preferred twice as much by Black consumers, these were a negligible part of the market despite signs of growth.

"Skin-whitening cosmetics go back thousands of years, to the ancient Mediterranean and Middle East, where wealthy women and men used white face powders to achieve a paler complexion."

According to the NHS, skin lightening, or bleaching, is "a cosmetic procedure that aims to lighten dark areas of skin or achieve a generally paler skin tone." One of the main ways to achieve this is by using skin-whitening products that aim to reduce the production of melanin, a pigment which gives our skin its colour and protects it from the sun.

Such cosmetics go back thousands of years, to the ancient Mediterranean and Middle East, where wealthy women and men used white face powders to achieve a paler complexion without blemishes. It signified high-class status, in contrast to those bound to the skin-darkening effects of outdoor manual labour.

***Blanc de Perle* (or "pearl white"), a cosmetic applied like modern foundation, c.1890.**

However, these powders contained substances that we now know to be toxic and harmful to human health. As early as 3000 BCE, elite ancient Egyptians made their iconic black eye make-up with kohl, whose main ingredients included galena, a lead sulphide mineral. Paired with white face paint that contained white lead, it produced dramatic as well as dangerous results. Ancient Greeks and Romans also used white lead-based powders to achieve pale and smooth skin. Further East, in China and Japan since the Qin dynasty (221–206 BCE) and Heian Era (c. 794–1192 CE) respectively, rich women and men remained indoors and employed white lead and rice powder cosmetics in pursuit of "flawless", untanned skin – a marker of wealth and nobility. In precolonial Africa, mineral and botanical skin care treatments were used to protect the skin from harsh sun and wind and achieve a glossier, lighter and brighter skin tone, considered a sign of elevated social standing and feminine beauty, and also of ritual and metaphysical significance. For example, those in states of ritual transformation – such as initiates, nursing mothers, healers and apprentice diviners – have covered their face in white clay widely across Africa for centuries to communicate their powerful connections to spirits and ancestors. In South Africa, white clay is still used today as a cosmetic and skin protector.

A tray of ancient Egyptian cosmetics. The ancient Egyptians were fond of dark eye make-up, used not just as a fashion statement but also to protect their eyes from dust.

"Social and gendered expectations have defined beauty ideas about white(r) and light(er) skin not only in Europe but around the world."

Although less damaging ingredients such as chalk, flour, rice, honey and olive oil were also used to make skin-lightening mixtures, white lead's ability to offer opaque coverage and stick to the face without an oil base made it a popular cosmetic ingredient in Europe from the 16th to 19th centuries. In the 16th century, Queen Elizabeth I used Venetian ceruse, a cosmetic based on white lead, to achieve her characteristically pale complexion. It is one of the earliest examples of celebrity cosmetic endorsements. However, the price she and her followers had to pay for a smooth complexion and the coverage of smallpox marks was skin discoloration, hair and teeth loss, and even death by lead poisoning.

Toxic ingredients continued to be present in cosmetics in the 1800s, with arsenic and mercury present in many off-the-shelf products advertised by magazines such as *Harper's Bazaar* and even endorsed by doctors. Arsenic complexion wafers – arsenic-

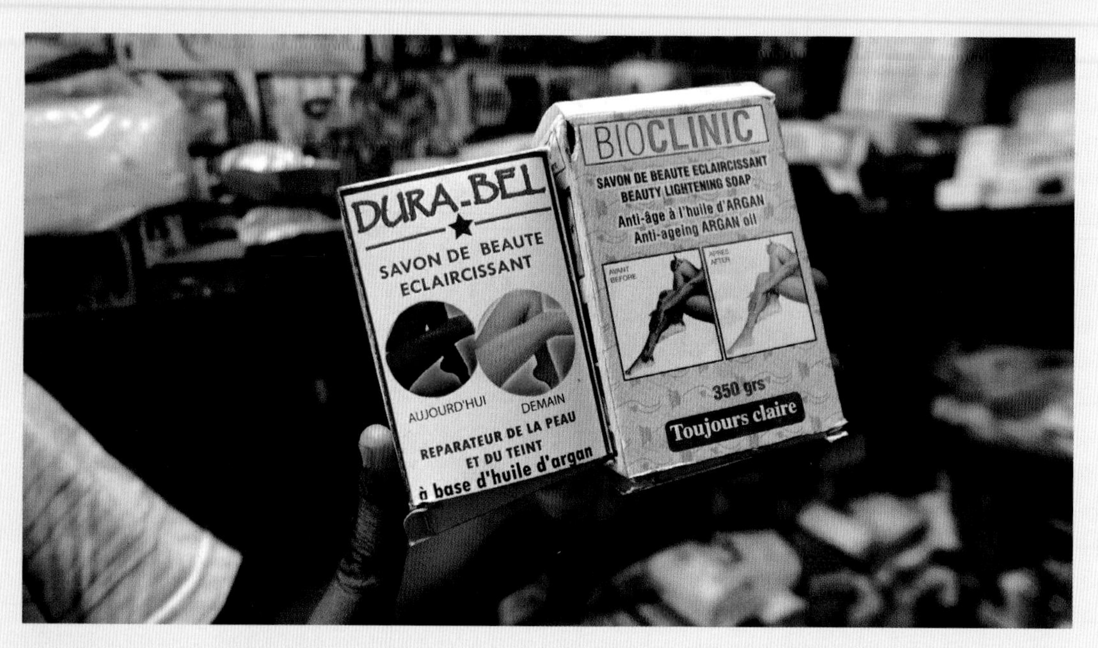

Skin lightening products in a shop in Ivory Coast.

filled tablets meant for indigestion – promised blemishless skin and were widely promoted as "harmless" from brands like British De Mackenzie's and American Doctor Campbell's in the 19th century. Not surprisingly then, British soap manufacturer Pears's *Blanc de Perle* contained bismuth oxychloride, a powder popular for its iridescence. Despite being non-toxic, bismuth oxychloride can lead to irritation and blocked pores if high concentrations of its crystalline structure are applied.

Pears, still in operation today under the Unilever brand, started selling what became the world's first mass-marketed transparent soap in 1807. Orchestrated by the advertising innovator Thomas Barratt, the company's marketing played a crucial role in its success. However, its adverts are infamous for racist connections between whiteness and "cleanliness". *Blanc de Perle*, meaning "pearl white", promoted similar ideas of whiteness. Marketed primarily to women, it also aligned with beauty ideals in 19th-century Britain and America linking femininity with virtue. "Virtuous" women had to look "natural" and "delicate", so homemade or shop-bought lotions were favoured over thick face paint, which was associated with sex and entertainment workers and considered vulgar.

Social and gendered expectations have defined beauty ideas about white(r) and light(er) skin not only in Europe but around the world. However, the European colonialists from the 15th century onwards disseminated a rhetoric of white supremacy using skin colour and physical differences as a way to exert control and justify exploitation and slavery. In the US, white enslavers would give preferential treatment to lighter-skinned enslaved people by letting them work indoors, for example. In India under British colonial rule, lighter-skinned Indians were deemed stronger and allowed to enter the colonial army. Discrimination that privileges lighter skin over darker is called "colourism", and it persists to this day, negatively impacting Black and Brown communities worldwide – including the public health threat from the use of harmful skin-whitening methods and products. While there are some cases in which skin-lightening products can be used to safely treat conditions related to irregularities in melanin production like hyperpigmentation or vitiligo, colourist ideas motivate many users to conform to white beauty ideals by using skin-whitening products, albeit shamefully and in secret because of the practice's negative links to colonialism and racism.

Colourism exists both between different racial and ethnic groups and sometimes within them too, where lighter skin is connected to beauty and physical attractiveness, wealth, success and power. In a systemically racist society, this can translate to better education, job opportunities and even health, as the majority of light-skinned Black and Brown actors, singers and politicians testify (think of the first Black American president). For women and girls of colour specifically, it's linked to better romantic and marriage prospects.

Since the 1960s, however, several campaigns have been proposing positive counter messages and images to the mainstream media's narrative that connects whiteness and beauty, starting with the "Black is Beautiful" movement by a group of Black creatives, including photographer Kwame Brathwaite. In the 21st century, charities such as the Beautywell Project and the Melanin Foundation are spreading awareness about the dangers of chemical skin-lightening practices, while wider justice movements such as Black Lives Matter have exposed beauty corporations' double standards of selling skin-whitening products while appearing to support racial justice. "Inclusive beauty", which aims to cater to all, regardless of gender, age, religion, race, skin tone and type, has become a household term since 2017, when pop singer Rihanna's beauty brand Fenty launched with 40 different foundation shades to suit a multitude of skin tones.

When it comes to health, regulations in cosmetics have become stricter in the last two centuries. However, global and local governmental regulations can't always keep up with a fast-growing market and e-commerce fuelled by increasing demand for skin-lightening cosmetics, particularly by expanding Asian and African middle classes. In 2013, the Minamata Convention on Mercury, a UN-backed treaty, pledged to "protect human health and the environment from the adverse effects of mercury". But cosmetic corporations and small companies alike continue to sell dangerous skin-lightening products, with ingredients that are often not listed on their packaging, especially mercury, as well as carcinogenic bleaching agents like hydroquinone, and steroids. Print, digital and social media and advertising have been the conduits of products, messages and images that promote whiteness as the route to a better appearance and life, a reflection of social inequalities and desires rooted in colonialism and racism. Although by a different company, a brightening skin lotion named *Blanc de Perle* is still sold today.

Despite beauty itself being subjective, shaped by continually changing social forces, its pursuit is a fairly universal human trait. Its applications, however, are historically and culturally specific: one's skin is its battleground.

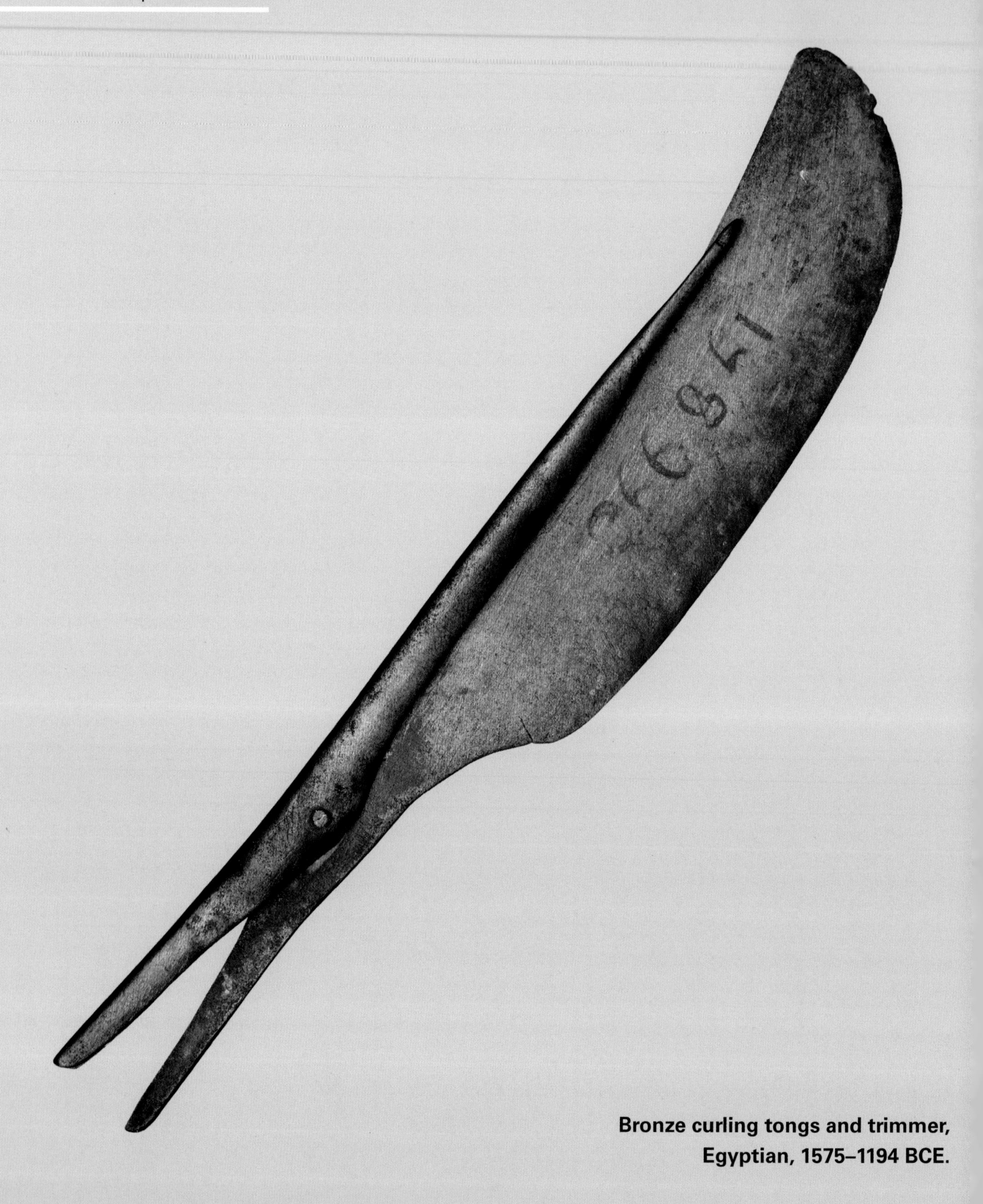

Bronze curling tongs and trimmer, Egyptian, 1575–1194 BCE.

HAIR TONGS

When it comes to fashion, curled hair is one of the few styles that has prevailed across time and different cultures. Today, heated curling irons are one of the most popular tools for hair and beauty care, both in professional salons and at home. In 2018, the global market for curling irons was valued at $3.22 billion, with demand for these products continuing to grow. But whether you're reaching for a BaByliss curling tong or a Dyson Airwrap, you might be surprised to hear that this cutting-edge hair curling technology actually has its roots in ancient civilisations, dating back thousands of years.

You only have to look at ancient artwork to see the demand for curls throughout history. Some of the earliest images of curled beards and head hair appear in Babylonian, Assyrian and Persian sculpture. Rags, pin curls, oils and adhesive products such as beeswax or resin were widely used in curling methods across cultures. But curling tongs stood out for their use of heat, producing the most long-lasting curls in the shortest time possible.

While we don't have an exact record of when humans started using tools to create the perfect curl, hair tongs have been discovered from as early as 1500 BCE. The ancient Egyptian hair tongs pictured opposite date from between 1575 and 1194 BCE. The Egyptians would heat these tongs over a fire, before winding individual sections of hair around one side and clamping it down with the handles to create a curl. Sometimes hair trimmers were attached, making these one of the earliest multipurpose cosmetic tools. Egyptologists largely agree that the elaborate curled, braided and decorated hair styles in sculptures and art throughout Egyptian history likely represent wigs made of human hair. These wigs were often designed specifically for ceremonial purposes rather than everyday wear, and being able to afford tools to curl them was a mark of high status.

"Some of the earliest images of curled beards and head hair appear in Babylonian, Assyrian and Persian sculpture."

Similar curling irons were also used by the ancient Greeks and ancient Romans from around 500 BCE all the way into the beginning of the 3rd century CE. Sculptures show us extremely elaborate curls, favoured by Roman women in particular, which towered like crowns above their heads. Curls emulated the looks of the goddess of beauty and love (Aphrodite for the Greeks, Venus for the Romans). The Roman *calamistrum*, or curling iron, was very popular

Persian and Mede dignitaries sporting curled beards, carved on the Apadana palace in Iran.

according to Cicero, and numerous examples of this tool have been found in Roman cities such as Pompeii.

The technology of hair tongs and how they were used remained largely unchanged until the 18th and 19th centuries. In 1765, a design emerged that had only one moving prong, similar to modern-day barrel-shaped curling irons. In 1866, Hiram Maxim, the American-British inventor responsible for the fully automatic machine gun and the mousetrap (to name but a few), obtained the first patent for a hair-curling iron. However, the most well-known innovator of heated hair tools was French stylist Marcel Grateau, who is widely acknowledged as the inventor of the modern curling tong in the 1870s.

Rather than using heated tools to produce tight ringlet curls, Grateau used his manual clamping hair tongs to create the "Marcel Wave", a hairstyle that consisted of levels of tight waves close to the head and usually applied to a bob haircut. A version more familiar to many people today is the "finger wave", a style that imitated the Marcel Wave by using fingers to form waves on gel-dampened hair that would need to set overnight. However, the Marcel Wave would not get much traction until French actress Jane Hading sported his styled bob decades later. Grateau was not granted a patent until 1905, and the Marcel Wave hit the height of popularity in the 1920s and '30s.

Despite these developments, the 19th-century versions of hair tongs still relied on being heated by fires, coals or gas burners, just like their ancient predecessors. Using metal heated with fire was dangerous for a multitude of reasons, and required a very skilled and steady hand to complete the process safely. The heated metal tongs would often burn hair or set it alight, and cause serious long-term damage to its quality. Stylists would have to place the hair between damp paper or cloth to avoid accidents.

It wasn't until the discovery of electricity that temperatures could be regulated more easily. While prototypes were developed as early as the 1920s, the first patented electrical version was produced by Rene Lelievre and Roger Lemoine in 1959, making the "Hollywood Curl", sported by beauty icons like Marilyn Monroe, available to a much larger market. However, these were often very inefficient, had single heat settings, and were tricky to use, requiring significant training in order to manipulate the clamps effectively and safely.

In 1980, Theora Stephens, a Black hairdresser, developed a curling iron with a spring closing clamp, heat control settings, and a more effective tool that did both the pressing and curling. Her designs were revolutionary, ultimately making the curling iron a household item, safe and easy to use. With the addition of a ceramic plate in the place of a metal one, the hair was less likely to become badly damaged from the heat. From the 1980s onwards, the hair curler became a staple of many hair care kits worldwide.

Today, curling irons and hairstyling tools are becoming increasingly multifunctional and inclusive. Hair irons now have automatic features to make them safer and easier to use, while energy-efficient multipurpose tools that simultaneously dry, smooth and curl are becoming available to people at home. Since the 2000s, the straightener has also been used as a multipurpose tool, both straightening and curling thanks to the improved quality of the plates.

The hairstyling industry has historically catered to individuals sporting traditionally Western, white hair types. Textured hair has long been overlooked in the production of heated curling irons, which can cause severe damage to curly or coarser hair types. Now, with the inclusion of automatic temperature controls and gently heating materials such as ceramics, titanium and tourmaline, electric hair curlers are more inclusive than ever before. Some curling irons also consider physical accessibility in their design. Lightweight, smaller versions of the curling iron, rotating air barrel curlers that require single hand use, and quick heating or cooling versions that require less time and effort in delivering a curl are all features that make hair tongs work better for people with physical differences.

From ancient roots to modern-day glamour, heated hair tongs have inspired beauty trends for thousands of years. If you've never tried one, make like the Egyptians and give it a curl.

Marcel Grateau, inventor of the styling technique of hairwaving, demonstrates how it is done.

MATCHES

The box of "Brymay" safety matches opposite was produced between 1936 and 1937 by a company called Bryant & May, based in Bow, East London. Together with lighters, matches can be found in most supermarkets and kitchen drawers and are one of the most common tools used to set things alight, be it candles, log fires or camping stoves. Making people's lives easier since the early 19th century, modern matches were one of the most revolutionary inventions in the history of fire-making tools. The path to their existence, however, was not only shaped by ingenuity and new discoveries, but also by the hard work of factory workers and the poor working conditions and injustices they faced. Sparking landmark industrial action and the dawn of a new trade unionism, these little boxes of wooden sticks brought about changes that went far beyond the ability to light a flame.

Their story starts at least one million years ago, where we have evidence for the extinct ancient human species *Homo erectus* using fire in a controlled way for the first time. It is assumed that our distant ancestors were inspired by naturally occurring flames from forest fires and lightning strikes. Realising that the heat they produced could be utilised, they began to look for ways to start their own. Since then, fire has been one of the essential foundations of human life. Throughout time, it has provided light in darkness, the warmth to survive in harsh environments, the means to process food, and protection from predators. It has made it possible to develop new tools and weapons, has been crucial for energy production, and is part of rituals and ceremonies in religions and communities around the world. Without it, human evolution would not have been possible in the way we know it.

"It is assumed that our distant ancestors were inspired by naturally occurring flames from forest fires and lightning strikes. Realising that the heat they produced could be utilised, they began to look for ways to start their own."

Throughout time, thinkers and makers from all over the world have grappled with the question of how to most efficiently create a reliable flame in the home, on the go, and in different environmental conditions, spanning humid tropical forests, windy mountains and rainy cities. It is unclear when exactly each method originated, but most have ancient roots and have been used all over the world in different regions, cultures and time periods. All methods work by increasing the temperature of tinder, which combusts, creates an ember and then heats up other

Matches in cardboard match box (with safety striker strip) by Bryant & May, 1936–1937.

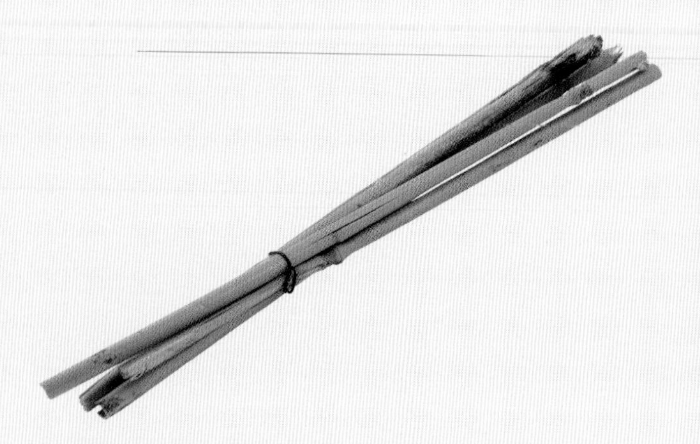

Prior to safety matches, wooden sticks dipped in sulphur were used.

material, called kindling, until it starts to burn as a flame. In percussion methods, a sharp-edged flint or hard stone is struck against a fire striker of mineral or fire-steel, which causes hot, oxidising metal particles to split off the fire striker and ignite the tinder. In friction methods, solid combustible materials are slowly ground against each other or a hard surface until their temperature is increased and an ember is produced. And in methods using the compression of air, like fire pistons, tinder is placed in a tube and a piston with an airtight seal is rapidly pushed into it. The air inside the cylinder is compressed and the pressure and temperature are increased until the tinder combusts.

Matches work through exothermal chemical reactions, meaning chemical reactions that release energy into their surroundings, which produces heat strong enough to set tinder on fire. The earliest known examples of matches originated in China around 577 CE and consisted of small sticks of pinewood coated in sulphur. These, however, were not self-igniting and had to be lit by an external source.

In the second half of the 17th century, European inventors began to experiment with different mixtures of self-igniting chemical substances. One of them was Henning Brandt, a German alchemist who discovered the flammable nature of the chemical element phosphorus. While many experiments created explosions, none of them were able to transfer flames safely onto slow-burning material like wood. The first self-igniting match that came close to success was produced in 1805 by French chemist Jean Chancel, who tipped wooden sticks with potassium chlorate, sugar and gum and ignited them by dipping them into an asbestos bottle infused with sulphuric acid. Due to the dangers of this process and the expense of the matches, they never became popular. In the following decades, several other inventors tried and patented different chemical compositions and igniting methods, with mixed results.

This changed when British pharmacist John Walker invented the first successful friction match in 1826 – allegedly by accidentally scratching a match on the hearth while preparing a lighting mixture. Coated with sulphur and tipped with a mixture of

"British pharmacist John Walker invented the first successful friction match in 1826 – allegedly by accidentally scratching a match on the hearth while preparing a lighting mixture."

antimony sulphide, chlorate of potash and gum, it was ignited by being drawn over a strip of sandpaper. These were dangerous, as flaming balls sometimes fell off the matches, and Walker never patented his invention. A few years later, an improved version called lucifer matches came on the market. The biggest change, however, happened when the French chemist Charles Sauria replaced antimony sulphide with white phosphorus, a more effective chemical substance, in 1830.

With this breakthrough, matchmaking became a popular business. One of the leading producers was Bryant & May, who established their first factory in Bow, East London, in 1861. By 1884, they were producing about 300,000 matches a day. Like most factory work, making matches was a tedious and poorly paid task, largely carried out by women and children from the poverty-stricken surrounding areas. This work was particularly dangerous. The poisonous nature of white phosphorus caused a condition called "phossy jaw", which destroyed the bones of the jaw, could lead to brain damage and was often fatal. On top of the health risks, the poor working conditions at the factory included staggeringly low wages, frequent fines for small misdemeanours such as talking or dirty feet, and abuse by the foremen.

In 1888, these poor working conditions triggered landmark industrial action now known as the "matchgirls' strike". Around 1,400 strikers, mostly women and teenage girls, walked out of the factory, supported by social activists like Annie Besant. The strike led to the growth of a new trade unionism, including the formation of the largest union for women and girls, and union organisation for unskilled workers, as well as shifts in the British Labour movement. Despite these

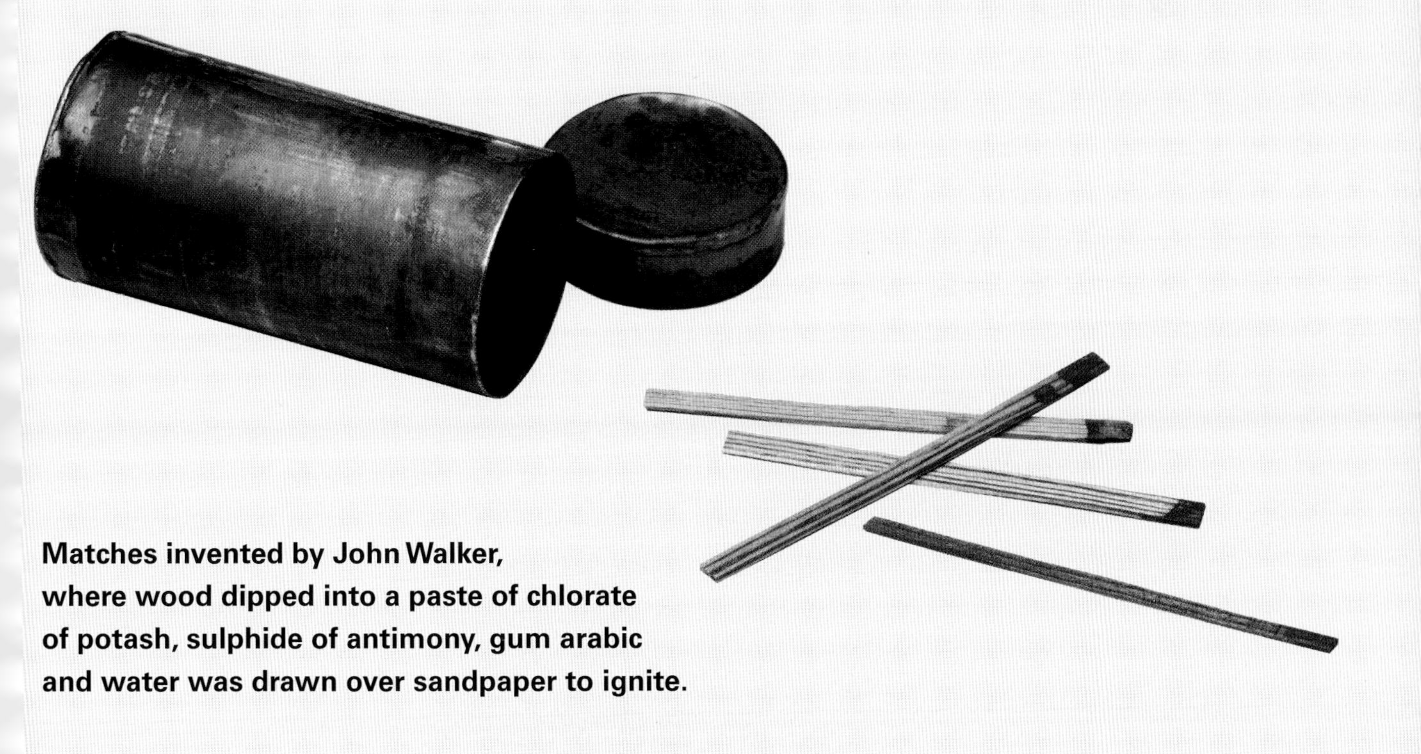

Matches invented by John Walker, where wood dipped into a paste of chlorate of potash, sulphide of antimony, gum arabic and water was drawn over sandpaper to ignite.

Match workers at the Bryant & May factory in Bow prior to their pioneering 1888 strike.

"The 'matchgirls' strike' led to the growth of a new trade unionism ... and shifts in the British Labour movement."

successes, and the company agreeing to implement small changes towards a better working environment, it took more than ten years for Bryant & May to stop the use of white phosphorus.

Eventually, the dangers of white phosphorus, which were known long before the matchgirls' strike, led to the development of "safety matches," like the box featured on page 31. This method was originally pioneered by the Swedish inventor Gustaf Erik Pasch in 1844, and Swedish companies long held a monopoly over their production. While matches coated with white phosphorus can be ignited on any kind of surface, safety matches work only on specially designed striking surfaces containing red phosphorus, which ignites due to its extreme reactivity with the potassium chlorate contained in the match heads. Today, safety matches have become the most popular type of match. In 1872, Finland, then part of the Russian Empire, was the first country to ban white phosphorus matches. In 1906, several other European countries followed suit after signing a treaty in Bern, Switzerland, at a conference of the International Association for Labour Legislation. It was only in 1908 that Britain passed its own White Phosphorus Prohibition Act, which was to come into effect two years later, more than 20 years after the women and girls taking part in the matchgirls' strike first raised awareness of the chemical's dangers. It is thanks to them that workers today are protected by UK labour law and legislations.

A lockout at Bryant and May's factory with peaceful demonstration, c. 1900.

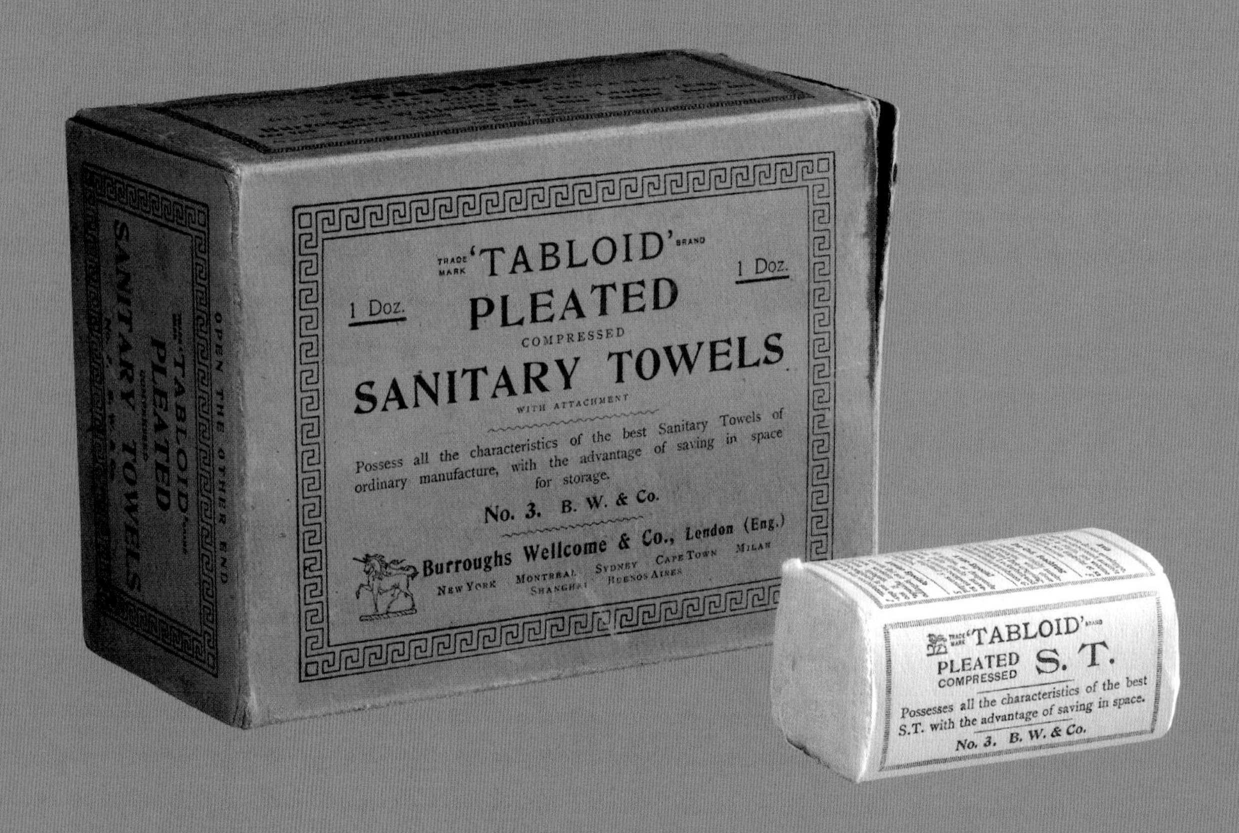

Twelve "Tabloid" pleated compressed sanitary towels with attachment, size number 3, by Burroughs Wellcome & Co., 1910–1940.

PERIOD PRODUCTS

Periods are nothing new, but you wouldn't think it from written records. Despite being a normal part of life for roughly half the world's population, and something that reoccurs month after month for many years, most historical sources do not deem menstruation worthy enough to document.

After Spanish conquistadors conquered the Aztec Empire in 1521, missionaries arrived to record the histories of the Mexica, an Indigenous people of the Valley of Mexico. They wrote about Aztec society, childhood, culinary practices and, particularly interesting to them, the Mexica practice of human sacrifice. Yet, in documenting a civilisation so presumably centred around blood, it seems the missionaries made an incredible omission in failing to record anything about how the Mexica dealt with their own blood every month.

"In documenting a civilisation so presumably centred around blood, it seems the missionaries made an incredible omission in failing to record anything about how the Mexica dealt with their own blood every month."

Nonetheless, the occasional menstrual clue prevails through the fog of history written by men. In the 4th century CE, Hypatia (the first female mathematician whose life and work is reasonably well recorded) is said to have thrown a used menstrual cloth at a man to get him to go away. This tells us two things: Hypatia wasn't afraid of standing up for herself, and the Romans most likely used pieces of cloth to soak up menstrual blood.

While periods are a constant thread running throughout human history, the way that periods have been dealt with has changed dramatically over the last few centuries. The use of cloths and free bleeding were universal ways to deal with menstruation around the world since periods began, and remained the main methods of period care for thousands of years. But this all changed in Victorian Britain, when menstrual products became more freely available.

All sorts of products have been used to deal with blood: tampons, cups, sanitary belts, period pants. But one of the most common products isn't so far removed from the cloths the Romans used. Today, period pads come in all sorts of shapes and sizes. Some are especially long for use during the night. Some can be washed and reused over and over, while others are made largely of cotton and plastic and stick to the gusset of underwear, ready to be thrown out after use.

But pads haven't always featured a handy sticky strip, and instead used to be held up with belts and buckles. A key figure involved in the creation of these sanitary belts is American inventor Mary Kenner. In the 1920s, Kenner designed a menstrual belt to solve common complaints that pads on the market were "too large, too long, too thick and too stiff". Today, a belt might sound like a cumbersome and awkward way to deal with periods, but Kenner's invention included easily adjustable straps to allow the wearer to comfortably alter the position, eliminating the chafing and irritation caused by most products on the market at the time.

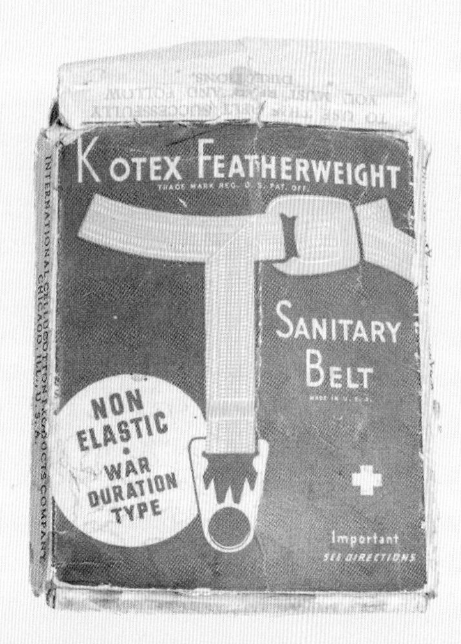

A sanitary belt like the one invented by Mary Kenner in the 1920s.

Kenner raised enough money for the expensive patent process in 1956. Numerous companies heard about Kenner's patent, but they all declined to work with her when they discovered she was Black. After her patent expired, manufacturers were able to profit off her idea, with Kenner never making any money from the belt herself. Despite the racist treatment she faced, Kenner filed a total of five patents in her lifetime, the greatest number of patents awarded to a Black woman by the US government. The revolutionary belts that she innovated were the main option in the West up until the 1970s and paved the way for the sanitary pad, yet they're largely forgotten about today.

During the 1990s and 2000s, there was a move towards disposable products that are used once and then flushed or thrown away. While these items shouldn't be flushed down the toilet, nearly half of the menstrual products used in the UK are, causing huge blockages in sewers as they join wet wipes, grease and food waste. These items are also found on beaches, and, being predominantly made of plastic, they float in oceans and languish in landfill sites.

Plastic is often combined with cotton in the production of disposable products. Organic cotton is generally better for the environment, nature and ecosystem, but even then, it is commonly used in conjunction with plastic, which comes with potential health concerns, as well as environmental ones.

During tests on certain period products, chemical agencies have reported finding

Menstrual cups are now a popular and more sustainable alternative to disposable period products.

various hazardous chemicals. Pesticide residues, plasticisers and fragrances are all commonly added to period products, and, unlike food and cosmetic manufacturers, period product producers are not required by law to state their ingredients. While these chemicals are often low in concentration, this feels like an important junction to question how these standards are set, and how transparent companies are with their consumers. However, there are alternatives to disposable plastic products. Over the last few years, there has been a dramatic increase in more sustainable options, such as organic cotton, reusable period pants and menstrual cups. While sustainable products are likely to be the future of menstruation, it's worth noting that reusable options tend to have a much bigger up-front cost even if they work out cheaper in the long run. It is difficult to have a sustainable period if you struggle to afford products in the first place. In 2022, ActionAid reported that as many as 12 per cent of users in the UK struggled to buy period products in the last six months, and as many as 500 million people worldwide don't have access to basic period products.

Before 2021, period products in the UK were classed as non-essential, luxury items, therefore invoking an additional 5 per cent sales tax. Despite this tax being scrapped, investigations are still being held to determine whether this reduction has helped consumers, or has been kept by retailers. Period products are a necessary purchase, and without tax cuts or the offer of free period products, those who can't afford them have had to turn to using cloths, toilet roll and socks. Period poverty has resulted in frequent absences from school, college or work.

Deciding which period products to use is up to each individual. There are already numerous social laws governing bodies and uteruses, and individuals must feel able to use whatever products feel most comfortable. And while progress is being made in the development of more sustainable products, more research is needed into the painful periods some people deal with, including premenstrual syndrome and diseases like endometriosis.

Compared to the vast sweep of human history, talking about periods openly, without shame or stigma, has happened only very recently, and only in some parts of the world. Hopefully, greater transparency about periods and the products available will ensure that people feel more confident and empowered when discussing their bodily functions, and ensure that they know what to look out for if something doesn't feel right and how to seek help if they need it.

ROLLER SKATES

Roller skating is a sport familiar to many people, conjuring images of childhood play, roller discos, roller derby or maybe even the dancing skaters whose talent has gone viral on social media over the last few years. However, this sport has a long history, as illustrated by the pair of Victorian skates opposite. From their inception in the 18th century, through the Victorian era to the American Civil Rights movement and well into the modern day, roller skates have facilitated significant social change.

One of the earliest records of roller skates comes from 1743, when they were used on stage in a London theatre to create the illusion of ice skating. However, the most memorable debut of roller skates occurred around 20 years later, thanks to the man credited with their creation – the Belgian-born inventor, and musical instrument maker, John Joseph Merlin. In the 1760s, after moving to London, Merlin debuted his new invention at a masquerade ball by entering the event wearing his roller skates and playing a violin. Unfortunately, due to difficulty steering and a lack of brakes, Merlin crashed into a mirror worth over £500, a value recorded in Thomas Busby's 1805 account of the incident. Merlin not only broke the mirror, but also his violin, and injured himself in the process. This was not necessarily the best advertisement for roller skating, and it wasn't until American inventor James Plimpton designed and patented his first quad skates in 1863, nearly 100 years later, that the sport took off in earnest.

"One of the earliest records of roller skates comes from 1743, when they were used on stage in a theatrical performance in London to create the illusion of ice skating."

Plimpton's "rocking" quad skates were much easier to steer and control. By incorporating a rubber cushion into the mechanism of the skate, the wearer was able to transfer their weight to the inside or outside of their foot and manoeuvre more freely. In 1866, Plimpton also opened the first roller rink in America in Rhode Island, where he offered roller skating lessons. This was a clever marketing tool and was influential in the growth of the roller skating trend in the US.

These new developments crossed the Atlantic in August 1865 when Plimpton submitted a patent in England for his improved skates. Many factors influenced the success of roller skating in the UK in the 19th century, particularly the Victorian passion for ice skating. However, at the time, this

These roller skates date from the late 19th century, when indoor roller rinks were fashionable.

"Roller rinks played an important role in the American Civil Rights movement, with numerous protests against segregation being held at these locations from the 1940s onwards."

was a highly weather-dependent and very dangerous sport because people skated outdoors on frozen bodies of water. In Regent's Park, London, in January 1867, the ice gave way and hundreds of people were submerged in freezing water, resulting in around 40 deaths. Although there were attempts to make indoor ice rinks prior to the development of artificial freezing technology, these rinks were made using a mixture of pig fat and salts, which gave them an unpleasant odour. Understandably, many Victorians were keen to find a safer and less fragrant alternative, which is exactly what roller skating provided. Towards the end of the 19th century there were a large number of indoor roller rinks across the country, many of which had in-house bands playing live music. The skates pictured on the previous page, dating from somewhere between 1870 and 1890, are an example of those popular during this time.

The Victorian fascination with roller skating, termed "rinkomania" by the press, went beyond the health benefits of the activity. One of the main attractions was that the sport was practised by both men and women. Courting couples were closely monitored during the Victorian era and roller skating provided the opportunity to escape the eagle eyes of their chaperones. This provided young people with the chance to socialise more freely, engage in prolonged physical contact and have private conversations. The rise of roller skating was also accompanied by developments in clothing, with many younger women opting for shorter, skate-friendly skirts.

While these aspects of the sport did not usher in a radical sexual revolution, they caused significant developments in Victorian society. These rather unassuming skates illustrate an important societal change, representing a sport that offered increased independence and empowerment for young people, especially women.

Roller skating's close ties to individual and collective empowerment continued into the 20th century. Roller rinks played an important role in the American Civil Rights movement, with numerous protests against segregation being held at these locations from the 1940s onwards. Into the 1960s, many rinks allowed Black skaters only one night a week with

Black roller skaters pack a rink on a Saturday night in Chicago.

clearly coded event titles such as "soul night" or "MLK night". Over time, various styles of skating developed among Black communities across the United States. Fuelled by variations in local popular music, different skate styles such as "JB-Style", "Trains & Trios", "Fast Backwards" and the "Slow Walk" emerged. These can still be seen today. Roller rinks continued to be significant spaces among Black communities into the 1980s and '90s. Many (now famous) hip-hop groups and artists such as N.W.A started out performing at roller skating rinks when they struggled to get bookings at other venues. However, these community spaces have struggled to survive in the 21st century. The 2018 documentary *United Skates* highlighted the prevailing African-American skate culture and those fighting against the widespread closure of rinks. Despite issues throughout the 20th and 21st centuries, the Black community has continued to skate, finding a sense of identity, inclusivity and freedom within the sport.

Ledger Smith, a semi-professional skater, is greeted by H. Carl Moultie, vice president of the D.C. branch of the NAACP, as he arrives by roller skates from Chicago to join the Great March on Washington, 28 August 1963.

The sense of inclusivity facilitated through skating was one of the great successes of the 1970s Roller Disco craze, which provided low-cost entertainment and was highly popular among Black and LGBTQ+ skaters. Roller discos provided spaces where individuals could go to express themselves among like-minded people. Even today there are many community groups worldwide that focus on bringing together Black, PoC and LGBTQ+ skaters, providing a sense of freedom and empowerment.

These attitudes also extend into the more formalised arena of roller skate sports, such as roller derby. Created in America by Leo Seltzer, roller derby began in 1935 with teams competing to complete 57,000 laps around a track. However, in 1937 this was redeveloped to become the full-contact, high-speed sport we see today. Played on an oval track, five players from each team skate at any one time and score points by having their "jammer" (point scorer) pass members of the other team. The other four members of the team (blockers) take on both offensive and defensive roles, using high contact techniques, such as body hits, to assist their jammer and stop the other team scoring points. From its early iterations roller derby has had strong feminist connotations. It is an early example of a female contact sport, and one that had equal rules for both male and female teams. By the 1970s the sport had gained huge popularity: over 27,000 people attended Shea Stadium in 1973 to watch the International Roller Derby League championship game. However, the league had disbanded by the end of 1973 and was sold on to Roller Games. This move was not welcomed by many because it was seen as relying too heavily on theatrics, and by the mid-1970s viewership had declined significantly. The sport continued in differing iterations into the 1990s, often featuring manufactured wrestling-style narratives, but these were unsuccessful in the long term. The sport's modern revival began in the early 2000s in Austin, Texas, and it is now practised by over 1,000 leagues worldwide. This 21st century resurgence was furthered by the presence of the sport in media and pop culture, exemplified by the 2009 film *Whip It*. Roller derby is a highly inclusive sport, welcoming players of all genders, sexualities, ethnicities, body types, ages and previous sporting experience. As a full-contact sport, roller derby encourages players to freely use their bodies and strength, creating an inclusive and empowering environment that represents strong feminist values.

Although it has been a long-standing part of many communities, roller skating underwent a recent resurgence in popular culture in 2020 during the Covid-19 lockdowns. Fuelled by people's desire to find new ways to keep their bodies and minds moving, as well as the viral success of skating videos on platforms such as TikTok and Instagram, many people took up this hobby. It was so popular that it even led to a worldwide shortage of roller skates. This sport allowed people to get outside and even safely socialise, enabling them to understand and experience the sense of freedom and liberation that roller skating has provided to individuals and marginalised communities throughout its history. This began a new iteration of the sport, not a revival, and the groups who have maintained a passion and love for roller skating throughout the years should continue to be acknowledged.

Throughout their history, roller skates have facilitated social change, community building and individual empowerment, solidifying the sport's position as more than a contemporary hobby. However, perhaps it's best to learn from John Joseph Merlin and steer clear of expensive mirrors and violins the next time you strap on a pair of skates!

"Roller skating underwent a recent resurgence in popular culture in 2020 during the Covid-19 lockdowns . . . It was so popular that it even led to a worldwide shortage of roller skates."

Roller derby has existed in various iterations since the 1930s and has long been considered an inclusive, feminist sport.

NYLON STOCKINGS

These two pairs of nylon stockings, one neatly folded in its cardboard box, the other carefully unwrapped, may seem unassuming at first. Probably a reminder of special occasions and elegant dress for some, but for others they doubtless trigger frustrated memories of accidental ladders and tears. The sheer, shimmering hosiery is, however, more than just that. Besides telling stories reaching from gender conventions to sustainability and repair culture, nylon stockings were a main protagonist in the events of World War II, and a major milestone in the development of new synthetic materials in the early 20th century.

The word *stocking*, defined as a "close-fitting garment covering the foot and lower leg", was first used in the 16th century and is related to the Old English word *stocu*, meaning "sleeve". The need to cover one's legs for warmth and protection, however, is much older than that. While in prehistoric times humans wrapped their legs in animal skins held up by leather belts or girdles, the oldest discovered socks were found in ancient Egyptian tombs. Socks and stockings are often referred to interchangeably, but for the sake of specificity, the latter are longer than socks and worn up to the knees rather than just covering the feet. A term that unites all garments worn directly on the feet and legs is "hosiery", deriving from the Anglo-Saxon *hosen*.

"While in prehistoric times humans wrapped their legs in animal skins held up by leather belts or girdles, the oldest discovered socks were found in ancient Egyptian tombs."

While in later centuries hosiery became mostly associated with women's fashion, the first hosen were almost exclusively worn by men. From the Middle Ages to the 17th century, they resembled close-fitting tights stretching from the waist to the feet. Thereafter, they were split into two parts: the "upper hose", also called "breeches", covering the body from the waist down to the knees, and the "nether hose", or stockings, covering the lower legs and feet. Not yet made of elastic materials but of wool, cotton, linen or – for more affluent wearers – silk, stockings were either tied around the legs with ribbons or held up by

Two pairs of velure hosiery, nylon fully fashioned stockings, 1950s.

A 1786 illustration showing a gentleman sporting silk stockings.

suspenders. Initially woven by hand, the first stocking frame knitting machine was invented by English clergyman William Lee in 1589, setting off a new era of mechanisation and mass production. By then, stockings had begun to permeate gender boundaries.

The first woman recorded to have worn a pair, made of knit silk, was Queen Elizabeth I in 1560. Flipping gender conventions around, stockings fell out of men's fashion in the mid-19th century, and were replaced by trousers. For women, on the other hand, they became an increasingly essential garment worn for warmth, protection and modesty. While skirts at first tended to be kept long and stockings short, rising hemlines throughout the 19th and 20th centuries placed more emphasis on the no longer hidden hosiery. By the 1920s, it became acceptable for women to show more of their legs, and the garments covering them became symbols of fashion and freedom.

The most significant change in the history of stockings began in 1928, when the American chemical conglomerate DuPont hired organic chemist Wallace H. Carothers to work at the company's Experimental Station. Having decided to fund research that was not immediately linked to the production of a product, DuPont gave Carothers the freedom to dedicate his time to any topic of his interest. Choosing polymer research, he began experimenting with the possibility of producing human-made synthetic materials by melting different organic compounds together. After years of unsuccessful attempts and disappointments, DuPont broke their promise of complete research freedom because they needed a commercially usable product.

Finally, the year 1935 heralded the dawn of a new synthetic age. Carothers managed to produce the first example of a synthetic polymer fibre by combining two carbon-based substances, which create a solid that can be stretched and made into a thread.

"By the 1920s, it became acceptable for women to show more of their legs, and the stockings covering them became symbols of fashion and freedom."

The new material was found to be stronger and more resistant to heat and water than any previous human-made fibre – a groundbreaking discovery. Today, we've come to know it as "nylon". Despite the new material's suitability for many different purposes, including toothbrushes, DuPont chose ladies' hosiery as the first market to explore. There is a common myth that the name "nylon" is a combination of New York and London, DuPont's two headquarters. In reality, however, its origin story is slightly less spectacular, and a little arbitrary. Initially, the fibre is said to have been called "norun", hinting at the fact that it was stronger than organic ones and that fabrics made from it would produce "no runs". Realising that this claim was not strictly true, DuPont's naming committee felt the need to correct it. After many unsatisfying attempts, they began to replace norun's letters one after the other until it felt like it rolled off the tongue, leading to the birth of "nylon".

At that time, American women bought on average eight pairs of stockings per year, mostly made of rayon or silk. While rayon relied on wood pulp, 90 per cent of silk imports came from Japan. This dependency was less than ideal in the years leading up to World War II, when the animosity between the two countries was at its height. Nylon,

The first experimental sale of nylon stockings in Delaware, USA, October 1939.

With stockings in short supply during World War II, many women took to painting their legs to give the illusion of hosiery.

however, required only coal, air and water, which were all readily available in the US.

The new material was considered a symbol of human superiority over nature, and a sign of hope in politically and economically challenging times. To stocking wearers, nylon promised to be cheaper, more durable, quicker drying and more elasticated. After making their first appearance at the 1939 World's Fair in New York, the stockings were released to the American public on 15 May 1940. Thousands rushed to the stores and, within four days, 4 million pairs had been sold. Shortly after, 30 per cent of the stocking market was dominated by DuPont, and nylon stockings had become an essential item in most American households.

A British World War II propaganda poster encouraging people to repair their clothes in aid of the war effort.

Only two years after their release, however, the blissful era of nylon stocking abundance came to a halt. In 1942, DuPont shifted its manufacture away from the consumer market to support the war effort. Later known as "the fibre that won the war", nylon began to be used for army equipment, such as parachutes, tire cords and mosquito nets, with the production of stockings ceasing entirely.

With the sudden lack of nylon stockings, which did not reappear on shelves until 1946, wearers had to care for and mend the ones they either already had or had bought on the black market. The latter flourished both in the US and in Britain, where nylon stockings were not officially available until after the war. Most stockings sold in the shops were made of rayon and silk, which were equally affected by wartime shortages. In Britain, a general "Make-Do and Mend" policy implemented by the government encouraged people – especially women – to mend and repurpose items, including stockings, to save raw goods. Mending shops provided repair services all over the country and, as a last resort, many women painted their legs with liquid make-up stockings, including black seam lines drawn with eye pencils.

"When hemlines continued to rise throughout the 1950s and miniskirts became popular in the 1960s, the suspenders used to keep stockings in place became inconvenient. Soon, the 'pantyhose' was invented, today mostly known as tights."

DuPont announced the restart of its stocking production in 1945, leading to customers rushing back to the shops and fighting to secure one of the desired pairs in what has now become known as the Nylon Riots. Soon, nylon stockings and their production spread across the world, with the pair shown on page 47 produced by the Swiss lingerie brand Triumph International in the 1950s.

A model wearing "Threads for Men", a range of modern nylon tights.

When hemlines continued to rise throughout the 1950s and miniskirts became popular in the 1960s, the suspenders used to keep stockings in place became inconvenient. Soon, the "pantyhose" was invented, today mostly known as tights. Within two years, tights had a 70 per cent market share and pushed stockings to the margins – almost going full circle to the close-fitting hose of the 16th century that covered the body from the waist down.

In today's tights, the less flexible nylon is often either mixed with or replaced by other synthetic fibres such as spandex, polyester or Lycra. After a few decades of endless availability of cheap mass-produced pairs, however, the climate crisis has moved sustainability back into focus. Due to most producers using crude oil to manufacture nylon, many consumers and brands have begun to choose either recycled or organic fibres instead.

Hosiery has throughout history been closely connected to different gender conventions, first as a staple of men's and later of women's fashion. This binary was, of course, never strictly adhered to by all parts of society, but today the blurred boundaries are becoming more widely accepted. After discovering that 50 per cent of its customers were men, the hosiery brand Threads created its new line "Threads for Men", which adapts its tights to fit all anatomies.

WHEELCHAIR

Wheelchairs are one of the most recognisable mobility aids, so much so that they are often used as a symbol for disability, indicating accessible toilets or parking spaces. Early forms of wheeled furniture date back thousands of years, and over the past 100 years there have been significant technological and design improvements. Even so, wheelchair users still face considerable barriers, and many parts of society remain inaccessible.

In the UK, there are more than a million wheelchair users, and many more people use wheelchairs temporarily for transport in places like hospitals. Wheelchairs are an essential tool for increasing mobility and independence, and are used for a variety of reasons. Not all users need one all the time – around a third of wheelchair users in the UK use them only occasionally, choosing to walk or use different mobility aids at other times.

Wheeled chairs, and other moveable furniture, have been used as mobility aids since ancient times. There are examples painted on ancient Greek pottery and engraved on stone in ancient China. In the 1500s, royals like Henry VIII (1491–1547) and Philip II of Spain (1527–1598) were pushed around in wheeled chairs when chronic illnesses like gout curtailed their movement. The first known self-propelled wheelchair was invented by a paralysed clockmaker in Germany in the 1600s. Using his clockmaking skills, Stephan Farfler (1633–1689) built a wheelchair that he could manoeuvre himself via a system of cranks and cogwheels.

"Wheelchairs are one of the most recognisable mobility aids, so much so that they are often used as a symbol for disability, indicating accessible toilets or parking spaces."

In the 1700s and 1800s, wheelchairs became a more familiar sight in the UK and were particularly associated with spas and bathing therapies. The healing waters of Bath, a city in South West England, made it a destination for many people with disabilities and health conditions. Visitors could rent wheelchairs to take them around town, and the most popular design of chair even became known

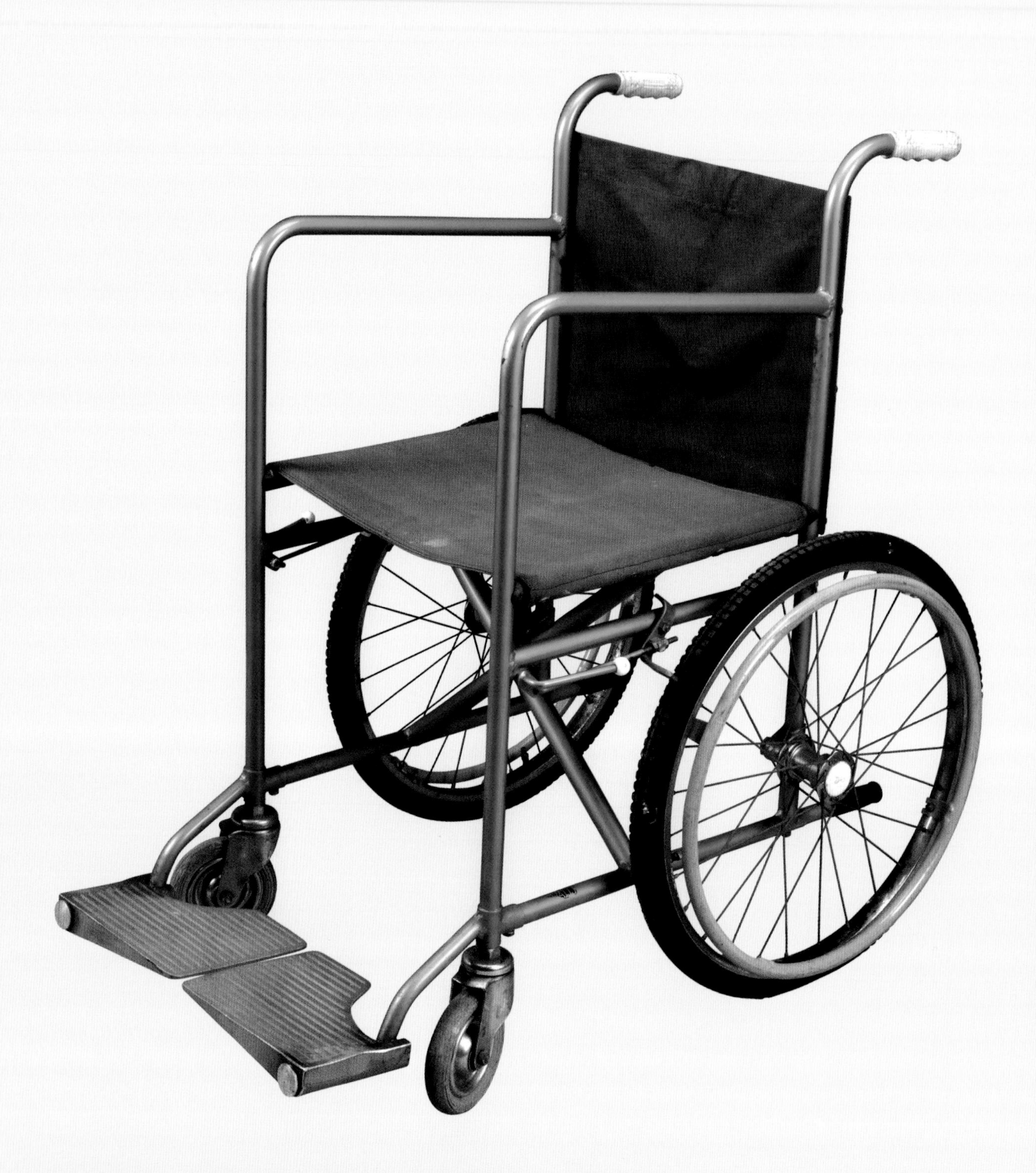

The Model 8, c. 1958, was a revolutionary design — push-rims on the wheels and an x-frame that was solid but flexible.

An ancient Greek urn depicting a wheeled, moveable chair.

as a "Bath chair" due to its association with the city.

The wheelchair pictured opposite is a Model 8, a very basic design from the 1950s, made by Dingwall. For many people, it would often have been their first introduction to wheelchairs. By 1961, over 60 per cent of wheelchairs prescribed by the NHS in the UK were Model 8s. Its simple, relatively lightweight design featured push-rims on the wheels that let people propel themselves, as well as handles that let others manoeuvre the chair. Between the wheels is an x-frame that enables the chair to fold, while keeping it rigid and stable when in use. Unlike previous folding wheelchairs, the x-frame also meant that the chair folded side to side rather than top to bottom. This was more convenient because it allowed the larger drive wheels to stay attached when folded.

The revolutionary design was first developed in the US in the 1930s, and was patented by two American engineers, Herbert Everest and Harry Jennings. Left paralysed after a mining accident in 1918, Everest asked his friend Jennings to help him design a wheelchair that would fit better in a car, enabling him to travel and commute more easily. The company they founded went on to become one of the first mass-producers of wheelchairs.

Despite the many improvements offered by this type of wheelchair, there were criticisms from users. One UK user complained that a prototype of the Model 8 had armrests that didn't fit under a table, and small caster wheels that didn't work well on carpet. Others found that it wasn't ideal for more active people; though its solid back wheels didn't puncture, they weren't well suited to outdoor use.

Although trialled in the 1910s, the first mass-produced electric wheelchairs were also developed in the 1950s, by Canadian inventor George Klein. Like many modern powerchairs, they were controlled using a joystick. In the UK in 1984, Dan Everard developed a powered wheelchair that could be used by children as young as 12 months old. This gave them the opportunity to develop their independence and explore the world around them like any other child.

Since the 1950s, the emergence of wheelchair sports and the growing needs of wheelchair athletes have contributed to the development of ultralight wheelchairs. Athletes have adapted their own chairs themselves to make them lighter, faster and more manoeuvrable. In the US, wheelchair basketball was particularly influential. Four-time gold-medal-winning US Paralympian David Kiley, for example,

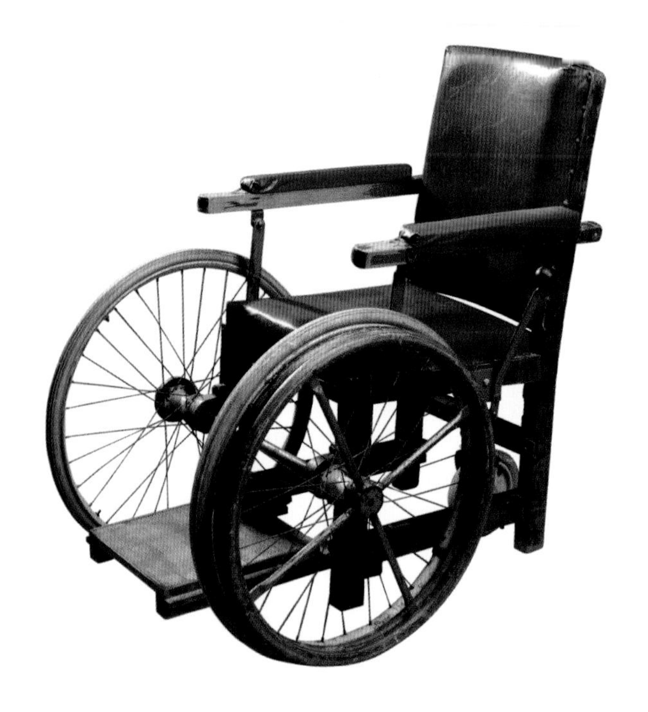

Child's wheelchair, Model 1, by Bencraft Ltd, English, 1940–1960.

pioneered adjustments to the centre of gravity of his wheelchairs, which made them much more responsive. Gradually, these types of improvements have made their way into mainstream wheelchair design, including using new composite materials, such as carbon fibre, rather than heavy metal structures.

Today, there are lots of varieties of wheelchairs available, from all-terrain and beach wheelchairs to folding powerchairs. They are also increasingly personalised to suit individual needs, personal preferences and expressions of identity. Personalisation can range from using a favourite colour to being custom-fit to increase comfort and support. However, access to wheelchairs can be unequal, as many are expensive, especially the latest and personalised models. In the UK, for example, there are limitations on the types of bespoke wheelchair that are available for free through the NHS. As with many technologies, the latest advances are often available only to those who can afford them.

Wheelchairs are vital for independence, yet the world around us isn't always accessible to wheelchair users. In the 1970s and '80s, disabled activists developed the social model of disability. This concept, still used today, highlights how people are disabled by societal barriers and discrimination, rather than by their impairments. One of the most contentious barriers for wheelchair users and other disabled people was, and remains, access to public transport.

Even into the 1990s, public transport in the UK was not required to be accessible. To travel on trains, wheelchair users were required to give three days' notice, and often had to travel in the guard's van with all the passengers' luggage. There was no legislation to prevent discrimination against disabled people, and many attempts to pass these laws had failed. Frustrated with years of hold ups, the Disabled People's Direct Action Network was formed in 1993. Through public protests, they brought attention to the inaccessibility of public transport. Their actions included bringing buses to a standstill on Westminster Bridge in London, and handcuffing themselves to trains at Cardiff Queen Street Station in Wales.

"One of the most contentious barriers for wheelchair users remains access to public transport."

Some members, who were arrested for protesting, were even freed because the police vans and stations were not accessible to wheelchair users.

Today, the law in the UK and across many countries states that all public transport must be accessible to all, regardless of disability. However, disabled people still find that their access is severely restricted. Buses have very limited wheelchair space, wheelchair users regularly report being stranded on trains because their pre-booked assistance hasn't arrived, and only a third of Tube stations in London have step-free access. This is partly because much of the infrastructure of the Tube is old, and only new lines, like the Elizabeth line, are more accessible. However, even more modern forms of transportation, like aeroplanes, are very inaccessible with airlines regularly damaging passengers' wheelchairs while transporting them.

The technology behind wheelchairs has moved forward hugely in recent decades, but the infrastructure needed to ensure that users are free to live and travel as they wish has lagged behind.

American athlete David Kiley, known for pioneering adjustments to his wheelchair to change its centre of gravity, plays in the 1988 Wheelchair Basketball World Championship.

FLUSHING TOILET

This toilet may look a little unusual, but it is still recognisable despite its unfamiliar flushing mechanism. The handle behind the seat is pulled to open a valve, which starts the flushing process. Named the "Optimus", it was designed in 1870 by Steven Hellyer, a British campaigner for better plumbing. The Optimus improved upon previous toilet designs, flushing more efficiently and quietly, but it was hard to keep clean.

As is the case with many groundbreaking inventions there was an evolution to achieve the most sanitary, quiet and efficient toilet design, but the toilet of today has remained largely unchanged throughout the lives of most living individuals, and many of us take our loos for granted! However, in the grand scheme of history the flushing toilet is a relatively new concept – so how did we get from cesspits and chamber pots to dealing with human waste during spaceflight within a century?

The problem of waste disposal is as old as time. Evidence of early waste disposal systems has been uncovered in archaeological sites across the world, such as drainage systems dating as early as 3000 BCE found in the Orkney islands, an archipelago in Scotland. Despite this early evidence, simple chamber pots and cesspits continued to be the most common choice of waste depository in the West until the late 1800s, so doing your business wasn't a particularly private or comfortable experience until relatively recently. That being said, if you could afford it, you could have a very detailed design on your personal chamber pot – and maybe even someone else to empty it for you!

"Simple chamber pots and cesspits continued to be the most common choice of waste depository in the West until the late 1800s."

It might be surprising that the first flushing loo in the UK dates back as early as the 1590s. In 1596, Sir John Harington, the godson of Queen Elizabeth I, published *The Metamorphosis of Ajax*, in which he described a flushing device that required 7.5 gallons (about 28 litres) of water to pump waste into the street or a water source. He even claimed that it could be used by up to 20 people before flushing! Harington had an Ajax flushing device installed at his home in Bath and later installed a working model for the Queen in Richmond Palace, but it clearly didn't catch on. This could be due to its high water requirement in a time before plumbing, as well as the lack of sewers.

The Optimus flushed efficiently and quietly — an improved design, even if it was hard to keep clean.

Until the 1850s, London's waste removal mainly consisted of cesspits beneath houses or ditches that led waste into the Thames. Many communities outside of London had their own independent drainage systems. Flushing toilets remained a luxury well into the 20th century, with non-flushing options such as outdoor "earth closets" especially popular in more rural areas. This type of toilet was patented in 1860 by Reverend Henry Moule, specifically to offer a more sanitary option to households without indoor piping. It wasn't until the engineering feat of modern sewers that the flushing of waste could truly be facilitated in a sanitary and efficient way.

Homeowners had been responsible for dealing with their own waste since a ruling of Henry VIII's in the late 1500s. But the British government's Public Health Act of 1848 transferred this responsibility into the care of local health boards, enacting that all new builds (and re-builds) should be equipped with a "means of drainage". This initiated a commitment to centralising responsibility for public sanitation, and marked the first stage in the creation of the sewer systems the UK so heavily relies on today.

However, the "means of drainage" put in place by local health boards continued to empty waste into water sources, many of which were used for washing and drinking. With the increasingly densely populated cities like London where flushing toilets were rising in popularity such improper drainage was a growing health hazard. This culminated in the "Great Stink" of 1858, when the smell got so bad that some MPs reportedly considered abandoning Westminster. It also led to increasing outbreaks of water-borne diseases such as cholera, as well as the belief that disease was spread through "miasma" (smells). The British government was forced to act.

A 16-year sanitation project was put into action that same year, with railway engineer Joseph Bazalgette at the helm. 1,100 miles (about 1,770 km) of drainage pipes were installed under the streets of London, feeding into 82 miles (about 130 km) of sewers that led to 4 pumping stations across the city via 6 "interceptor" sewers. It was a huge feat of engineering that funnelled waste 8 miles downstream of the city, into the water sources of less populated areas. Bazalgette's brick-lined sewers remain the foundation of today's sewer system. In the midst of this sanitary revolution in 1851, George Jennings installed the first flushing public loos – or as he named them, "monkey closets" – at the Great Exhibition in Hyde Park, London. Known as a pioneer of sanitary science, Jennings also contributed to improving the flushing lavatory. His "wash-out" design improvement in 1852 remained one of the most popular toilet designs throughout the century. By the 1870s, the sanitary business was booming.

A common myth is that Thomas Crapper invented the toilet, also coining the slang term "crap". However, he was a contributor to the popularisation of the flushing loo more than he was an inventor. His fame is possibly due to his questionable advertising practices – implying the invention of the "Silent Valveless Water Waste Preventer" was his own, for example. This was in fact invented in the early 1800s (patented in 1819) by Albert Giblin. Crapper took this design more than half a century later and popularised it, succeeding in taking the credit too.

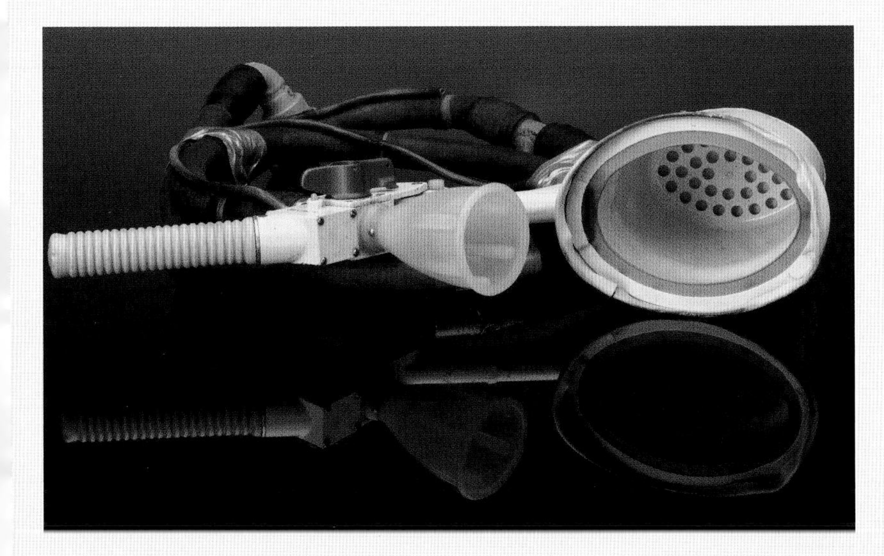

The Soyuz space toilet, (c. 1970s), with white enamelled metal collection unit for faeces and moulded plastic nozzle for urine, two suction tubes and outlet attachment.

Crapper did, however, invent a handful of related toilet add-ons, most notably an adaptation of a portion of pipe leading out of the toilet bowl known as an "S-pipe" (due to its "S" shape) that traps water after flushing to prevent odours and toxic gases escaping out of the toilet bowl. He turned this into a "U" shaped pipe, known as a "U-bend", which is still the Western standard to this day as it more efficiently traps water, is not prone to drying out and does not jam. Crapper held nine patents, but none related to the flushing toilet itself. His connection with the phrase "crapper" was coincidental, as "crap" had been used in relation to waste long before his lifetime. It is likely that the connection was strengthened by the size and success of his company, with "Crapper" branding emblazoned across all his products.

Although toilets started to be moved inside UK homes from around the 1920s, it was still the norm to have an outside loo well into the 1960s. Your grandparents may have stories about loos in their back garden as a child! With this in mind, it makes it all the more impressive that by the 1970s there were toilets adapted to work in the low gravity conditions of space – with the Soviet Union's Soyuz capsule leading the way in 1967, followed by NASA's "Waste Collection System", first used on the Skylab space station in 1973. These toilets utilised suction and the flow of air instead of water, and the current toilets in use on the International Space Station are modern adaptations of the 1967 Soviet design. The alternative for astronauts was most like a diaper and perhaps surprisingly, diapers are still used during spacewalks, but only as a "just in case", since astronauts understandably opt for a seat inside instead!

While toilets have remained relatively unchanged in the West, sanitation practices differ around the world, with squat toilets common in many countries, and bidets popular elsewhere. It is estimated by UNICEF that around 60 per cent of the planet – 4.5 billion people (as of 2019) – either do not have access to safe sanitation or do not have a toilet in their home.

"A photograph of a young woman" by Clementina, Lady Hawarden, around 1862.

SELFIES

The first online instance of the term *selfie* is widely credited to Australian internet user Nathan Hope. In 2002, he posted a photo from his birthday night out on social media and captioned it "... sorry about the focus, it was a selfie". Already common slang in some circles at the time, the colloquial term soon spread across the world.

Selfie is a hypocorism (a shortening of a name to show affection) of *self-portrait*. In 2013, the word was officially added to the Oxford Dictionary online, defined as "a photograph that one has taken of oneself, typically one taken with a smartphone or webcam and uploaded to a social media website".

The ubiquity of selfies across generations, cultures and nations has made it a social phenomenon. The selfie is about oneself, and the act of selfie-taking is a social ritual. While an image taken of oneself by oneself might seem narcissistic, selfies are designed to be shared with others for their amusement or approval.

If a picture is worth a thousand words, a selfie is a perfect example of how we make our own identity. A single image can communicate multiple, complex things about a person to peers, friends, family and social groups. Selfies empower us to promote who we are and to guide how people see us. The content and style indicate our interests and signify our similarities and affinities. In this way, the act of taking (and sharing) a selfie is to create a sense of belonging.

"Selfies empower us to promote who we are and to guide how people see us ... The act of taking (and sharing) a selfie is to create a sense of belonging."

In 2016, University College London (UCL) identified multiple selfie "genres" during a global anthropological research project on the uses and effects of social media. At the time of the study, they identified three main genres among school children in the UK: "classic" selfies featuring just one person, "groupies" with friends, and "uglies" depicting people at deliberately unattractive angles. While the terms for these genres may have changed, these categories remain recognisable. Since UCL's study in 2016, other global trends have emerged, such as adding cartoon filters over selfies (initially popularised by the app Snapchat), mirror selfies and the 0.5 selfie, where users take a photo from above in order to depict themselves with super long arms and tiny bodies.

Despite selfies seeming a relatively new phenomenon, they have actually been a part of the photographic canon since its inception. These "original" selfies were made for the same reasons as their modern-day counterparts: for self-expression, to tell a story about the taker, signifying and reinforcing their identity, and to share these images with others in their social group for approval.

The earliest selfie to be taken with a camera predates the term by 175 years. This selfie is attributed to Robert Cornelius in 1839. The photograph is of Cornelius from head to waist, arms crossed, head tilted at the camera, gaze steady. On the back of it is written, "The first light picture ever taken". Quite a claim. And in some ways, it is this description, not its "classic" selfie style as defined by UCL, that makes this photograph a selfie. It is posturing. Cornelius captured himself being the first to do something and then shared it for "likes" (or the equivalent acclaim in 1839).

The most shared selfie in Twitter's history, posted by Ellen DeGeneres in 2014, accumulated over 3.4 million retweets and 2.4 million likes.

Another early example of a Victorian selfie is shown on page 62, "A photograph of a young woman", taken by the woman herself, Clementina, Lady Hawarden. Like Cornelius's, this image ticks off the required tropes to make it worthy of posting to "the grid".

Through this photograph, Lady Hawarden conveys many messages about herself to the viewer. Note her direct, defiant look; her voluminous dress, which takes up almost the whole frame; her skill at taking a striking self-image. All of this promotes a powerful standing through an image, made by a woman, at a time when society had little space for women.

The internet is full of articles detailing how to take the best selfie, with tips including pointing your chin slightly down, looking up at the camera and holding the camera to the side. All of these contemporary tips appear strangely applicable to Hawarden's photo from 160-odd years ago. Perhaps she can be thought of as the instigator of the perfect selfie. The only difference: Victorians did not smile in their photographs. The "no smile" rule was on account of the long time it took for the photograph to be taken (that is, for the image to be exposed onto the photographic plate). For some camera models, this was up to 15 minutes – far too long to hold a toothy grin or pout. It was this long exposure time that allowed for self-portraiture, long before the selfie stick and smartphone front camera. The photographers could set up the scene, pull up the shutter, quickly step into position, and then freeze as the photograph was exposed. However, Hawarden did not have this long to strike her pose. She used the wet collodion process instead of the daguerreotype, which dramatically reduced the exposure time. The wet collodion process was a relatively new technique in photography for Hawarden's time, suggesting she was in step with photographic innovations. The collodion technique is where a photographic plate is painted in collodion (a mixture of nitrocellulose, alcohol, iodide and bromide), then dipped in silver nitrate (which bonds with the iodide and bromide to make light sensitive silver halide). The plate is added to the camera before it dries. All of this happens in a dark room. Once the camera is in place and photographic composition arranged in front of the lens, the shutter is removed, allowing the scene to be exposed onto the plate. The plate is then treated with a developer solution, washed, varnished and, finally, printed.

Hawarden was and continues to be considered an amateur photographer. Despite being an exhibited artist and having had her images showcased in the annual Photographic Society of London in January 1863, she was never paid for her photographic work. The Victorians deemed it inappropriate for high society ladies to be treated as professionals and earn money from their craft. Regardless, Hawarden was a prolific photographer and a pioneer of the medium. She took more than 800 images during her short career, using up to seven different types of camera. Her use of natural light, playful shadows and mirrors to create "double bodies" was deemed ground-breaking, as was her collodion technique.

Hawarden was made a member of the Photographic Society of London after exhibiting there. Her photographs, which primarily captured her adolescent daughters, are considered to offer rare glimpses (rare for the time Hawarden was working in) of women on the cusp of adulthood, making them radical pieces. They are hailed for their intimacy and subtle sensuality. In turn, the setting of Hawarden's images, almost solely in the first-floor studio of her London home, becomes emblematic of Victorian women being unable to "own" spaces outside of the home. As such, Hawarden's photographs tell stories both within and outside of their frames.

Hawarden's selfie is thought to be one of the few images she took of herself and one of the few that exists of her at all. This scarcity suggests a person shy in front of the camera unless she is in control of the lens. This shyness might be why some people take selfies today – to give them more control of the identity they create and promote. In Hawarden's selfie, there is a strong sense of defiance and identity. Much of her work remained in family photo albums, for the private viewing of the family. Perhaps Hawarden wanted to create a strong identity as a role model for her daughters.

Taking selfies is a social behaviour, and selfies were important to us long before the term existed. The rapid ubiquity with which selfies have become a mainstay in our lives only further proves their social importance. While the popular phrase "Pics or it didn't happen" may not be true, selfies do help define who we are.

THIMBLES

The distance between the west coast of Ireland and the east coast of North America is more than 5,000 miles (about 8,046 km). At its deepest, the Atlantic Ocean is 5 miles (8 km) deep. Storms are common, with waves reaching up to 60 ft (about 18 m) high. In the mid-19th century, it could have taken a ship up to 12 weeks to travel from Ireland to North America, depending on the weather. Now, commercial vessels make the journey in 3 to 4 weeks, and aeroplanes take 7 hours. How, then, did a simple, silver thimble provide the impetus for a message to reach from Waterville in Ireland to Nova Scotia in Canada in mere seconds on 12 September 1866?

Thimbles originated as a practical invention to protect a person's thumb while sewing. The first to be made in Britain was created in 1695 by a Dutch metalworker named John Lofting. Collecting thimbles became popular after special thimbles were made to commemorate the Great Exhibition in Hyde Park, London, in 1851. Over the centuries, they've even been turned into decorative objects, with some covered in the finest jewels. However, as demonstrated by this particular transatlantic episode, they also have other, somewhat unexpected applications.

At the beginning of the 1840s, conveying a message from one side of the Atlantic to the was only possible by sending a letter by ship. However, this period was one of rapid technological advancement, and one of the most important advancements was the sending of the first telegraph message in 1844. Developed by Samuel Morse and other inventors, the telegraph machine revolutionised personal communication. The resulting machine produced a single-circuit telegraph that worked when the operator pushed down a key to complete the electric circuit of the battery. This action sent the electric signal across a wire to a receiver at the other end.

"Thimbles originated as a practical invention to protect a person's thumb while sewing, but they also have other, somewhat unexpected applications."

The system was simple and needed only a key, a battery, a wire and a receiver. Unlike the mobile phones that we use today, this method was dependent on wires or cables being able to link two destinations. In order for a message to reach somewhere on the other side of an ocean, a cable needed to reach across it. As the potential of such a system was endless, progress came at a rapid rate. An underwater cable was laid between the UK and France in 1850. Linking the UK and America was next on the agenda.

This is the thimble that demonstrated the success of the first attempt to repair a transatlantic cable, c.1866.

There were several attempts at establishing a successful transatlantic cable. The first began in 1854 and was completed in 1858. The cable enabled Queen Victoria to send a message to the then President of the United States, James Buchanan. The cable, from Valentia Island in Ireland to Heart's Content, Newfoundland, would last just three weeks before breaking.

A second attempt took place in 1865, and involved a ship designed by civil engineer Isambard Kingdom Brunel. The SS *Great Eastern* was, at the time, the largest ship ever built, and had been converted to a cable-laying ship from a passenger carrier. Halfway across the Atlantic, the cable snapped, and the attempt was abandoned, with the cables left in two pieces in the ocean.

The third, and finally successful, attempt began in 1866 in the Anglo-American Cable House on the Telegraph Field in Foilhomurrum Bay on the west coast of Ireland. *Great Eastern* steamed westward across the Atlantic, lowering the cable onto the ocean floor as it went, and came ashore at Heart's Content on 20 July 1866. The mission went to plan, and the cable was put into operation.

On the journey to place this specific cable, *Great Eastern* also managed to find and repair the damaged version from the previous attempt. Two operational cables opened up the possibility of creating a loop. Latimer Clark, an engineer for the Anglo-American Telegraph Company, asked operators at the Heart's Content Cable Station in Newfoundland to connect the 1865 and 1866 cables to form a single circuit spanning 4,000 miles (6,440 km)across the Atlantic and back. To ensure it was operational, the newly created loop needed to be tested.

"Clark borrowed a thimble from Emily Fitzgerald, the daughter of a wealthy local landowner, and added a few drops of acid and a slip of zinc to it."

So, Clark borrowed a silver thimble from Emily Fitzgerald, the daughter of a wealthy local landowner who just happened to be visiting, and added a few drops of acid and a slip of zinc to it. This created a simple voltaic cell battery with a small amount of charge. He connected this battery to the underwater cables and then to a galvanometer, an instrument used to detect electric current, which showed that a very small electric current was indeed present within the cable loop. This demonstrated that the current created by the thimble, acid and zinc had travelled under the Atlantic to Canada, and then all the way back to Ireland.

The thimble was presented to the Science Museum in 1949 by one of Emily Fitzgerald's nephews. In the journey from primitive signals to satellites, the laying of the transatlantic cable, and its first messages, were undoubtedly an important step. Telegraph messages allowed for almost instant communication over huge distances for the first time in human history, meaning that everything from politics to people's love lives could eventually travel around the world. The small yet mighty thimble was an integral part of that incredible journey.

An 1865 illustration of workers on board the *Great Eastern* laying the first transatlantic cable.

MOBILE PHONE

On New Year's Eve 1984, Michael Harrison left a family party in Surrey and drove to Parliament Square to make the first ever call on a UK commercial mobile network. Almost 40 years after that momentous call, the UK – along with the wider world – has undergone a revolution in telecommunications, which has enabled mobile phones to evolve from a luxury item to an essential facet of modern-day life.

The earliest mobile phones began as radio communication devices and served primarily as army and policing phones. In 1895, inventor Guglielmo Marconi developed wireless radio telegraphy and a year later patented the first radio wave-based communication system in Britain. However, this technology was mainly designed for use in the maritime industry and by the Royal Navy. In the 1940s, private mobile radio services were introduced in the UK, which were mounted in cars. However, private mobile radio systems were limited by the fact that they could not be connected to the public telephone network, preventing mobile users from making or receiving conventional calls. There were also a limited number of channels and only a pair of people could use one channel at the same time. Despite this, the service quickly gained popularity among businesses and wealthy individuals during this period, although its high cost meant it was out of reach for most.

It wasn't until 1947 that the concept of a cellular network was proposed by Bell Telephone Laboratories in the US. Unlike a private mobile radio system, a cellular network divides the country into small geographic areas called "cells", and uses a low-power radio base station to enable communication between mobiles within that cell.

"The earliest mobile phones began as radio communication devices and served primarily as army and policing phones."

In the 1970s, researchers at Bell Labs continued to experiment with the concept of a cellular phone network. They aimed to cover the country with a network of hexagonal cells, each containing a base station. Messages from mobile phones were both sent and received by these base stations over radio frequencies. Any two cells that touched would operate at different frequencies to prevent interference. The base stations connected the radio signals with the main telecommunications network, and the phones switched frequencies as they moved between

With four different colour panels, the Ericsson GA 628 was the first mobile phone that could be easily customised.

Italian inventor Guglielmo Marconi (left) and his assistant George S. Kemp receive transatlantic wireless signals, early 1900s.

one cell and another. By the end of the decade, this cellular network was being rolled out by the Bell Labs Advance Mobile Phone System (AMPS) on a small scale. Today, AMPS is referred to as a "first-generation" cellular network, or 1G.

The first-ever call on the cellular network was made on 3 April 1973, when Motorola engineer Martin Cooper phoned Joel Engel of Bell Labs using a handheld prototype phone called the DynaTAC. This moment is generally considered to be the birth of the modern mobile phone.

Motorola continued to develop this prototype until 1984, when they launched the first commercially available handheld mobile phone – the DynaTAC 8000X. However, at roughly 25 cm (10 in.) long and weighing in at around 1 kg (2 lbs), it would be unwieldy by today's standards, and was nicknamed the "brick phone" by users even then. Priced at around £3,315 – almost £8,630 today – it was also inaccessible to most, and was mainly used by wealthy financiers and entrepreneurs.

The transition to GSM (Global System for Mobile Communications) in the early 1990s was significant in creating a mass market for mobile phones. Unlike the first cellular networks, which used analogue signals, GSM systems would transmit digitally. They were known as "second generation", or 2G systems. This transition enabled more networks to enter the market, which in turn drove handset prices down and made pay-as-you-go contracts more affordable.

The miniaturisation of mobiles phones also became essential to their growing popularity by the end of the 1990s, especially among

"IBM's Simon Personal Communicator is one of the first versions of what we'd call a smartphone."

teenagers. As they became increasingly practical, portable and accessible to young people, companies began to provide opportunities for phone customisation. The Ericsson GA 628, first released in 1997, was the first mobile phone to be easily customised, and provided four different front panels in green, yellow, blue or red.

Nokia also embraced the potential to personalise mobile phones and market them to a younger audience. In 1997, the company launched the Nokia 6110, which was the first Nokia phone to have the game *Snake* pre-installed. This was a milestone moment for the mobile gaming industry, and there are now more than 420 *Snake*-like games on iOS alone.

In 1999, Nokia released their 3210 model. One of the bestselling phones of all time, the Nokia 3210 allowed for customisation with interchangeable Xpress-on cover cases in different colours, which could be fitted over the phone. These could also be replaced without invalidating the phone's guarantee.

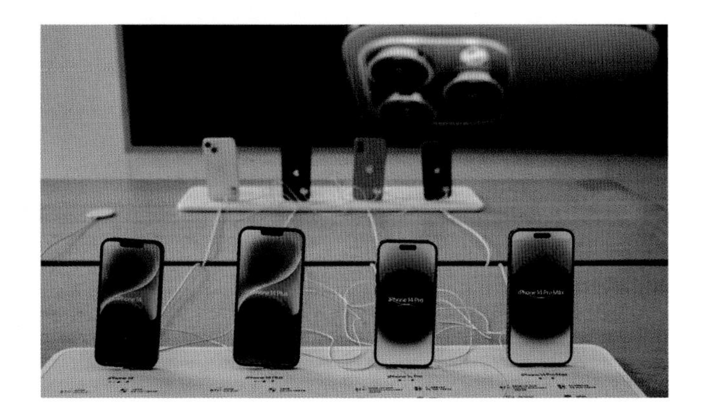

The iPhone 14, released in 2022.

In the early 2000s, phones such as the Sidekick facilitated the use of mobile phones as social communication devices. First released in 2002, the Sidekick debuted under carrier T-Mobile and popularised the concept of mobile internet. It featured an always-on GPRS data connection that allowed users to access email and instant messaging services like AOL Messenger.

As mobile phone ownership increased, so did the development of smartphones. Based on its functionality, IBM's Simon Personal Communicator is one of the first versions of what we'd call a smartphone. Released in 1994, it featured a touchscreen interface and stylus for touch input. However, many consider that the first Apple iPhone, launched in 2007, is the first fully realised smartphone. The iPhone was introduced by Steve Jobs, one of Apple Inc.'s founders, at a Macworld keynote in 2007. The phone allowed for a full internet experience, like a laptop or desktop computer, and featured multi-touch gestures, an iPod music and video player app, a dedicated YouTube app and a maps app powered by Google Maps. Despite these features, the iPhone was still limited by the fact it had no high-speed data access or third-party apps.

This changed a year later with the launch of its successor – the iPhone 3G – in 2008. It arrived just at the right moment, as 3G ("third generation") networks were becoming established, which provided faster wireless access to the internet and features such as

"Whatever shape mobile phones take in upcoming years, they may become even more indispensable to everyday life."

video calling. The iPhone 3G facilitated the use of these features, as it allowed for internet browsing on a more compact device for the first time. It also introduced Apple's App Store for the first time, where users were able to install third-party apps.

The iPhone set the precedent for every smartphone that followed, with competitors quickly launching rival products. Several competitors, including Nokia, were sceptical that Apple could compete in the mobile phone market as the company had previously been known only for their Macs and iPods. However, with one in five people worldwide owning a smartphone by the end of 2013, the iPhone had a profound influence in the growth of the smartphone industry.

Nokia's resistance to the smartphone evolution ultimately led to their phones' declining popularity, and they left the smartphone market in 2014 when their mobile phone business was purchased by Microsoft.

In 2012, the UK's first 4G service was launched by the British mobile network operator EE, which took download speeds up to 12mbps and greatly improved video streaming and video calling. As a result, smartphone screen sizes continued to grow to maximise the performance of these features.

As of 2023, 86 per cent of the global population owns a smartphone. Today's mobile phone is much more closely embedded in our day-to-day lives than ever before, especially with the growth of the Internet of Things (IoT). This is used to describe the growing network of internet-enabled devices that use AI technology, including sensors and actuators, to inform us about the status of everyday items while also allowing us to interact with them.

The IoT has brought us innovations like fitness trackers, smart thermostats and virtual assistants. Its use was accelerated by the outbreak of Covid-19 in 2020 to develop contact tracing devices and health-monitoring wearables, which provided crucial data to help fight the virus.

What's more, the launch of the 5G network in 2019 has provided us with faster network and download speeds, improving access to things such as remote healthcare, public services and entertainment. This increase in speed has allowed IoT devices to communicate and share data much faster than before.

Almost 40 years after the release of the DynaTAC 8000X, mobile phones have developed from a basic tool used to make and receive calls to a versatile pocket computer and personal device that has revolutionised communication, enabling us to stay in constant contact through calls, texts, social media, the internet and apps.

Whatever shape mobile phones take in upcoming years, they may become even more indispensable to everyday life as the IoT continues to develop, and it will be exciting to see what role mobile phones will have in our lives in the future.

OFFICE CHAIR

Movement, adjustability and good support are the elements that make up an office chair. All chairs are designed to support us when seated, which we have been doing with increasing frequency in recent years. Industrialisation, the rise of the information age and the integration of technology into our everyday lives have led to a more sedentary lifestyle. More office-based jobs require equipment that keeps professionals comfortable (and therefore more productive) during long periods of sitting at a desk. As such, scientists and designers have been striving to make the perfect office chair, combining functionality and safety with aesthetics and, most recently, sustainability.

The humble chair itself, though, dates back thousands of years, as murals, vases and frescos from ancient China, Greece and Rome show. The oldest surviving four-legged chair in the world is ancient Egyptian. The office chair has a longer history than we think, dating back to the 19th century and to the scientific advancements of the time. In the 1840s, the naturalist Charles Darwin created a proto-office chair by adding wheels to his armchair to travel faster to his specimens. However, only a small, privileged number of people (such as royalty and intellectual elites) owned those early chairs. Sitting on a chair was a declaration of power and higher status: you were setting yourself apart from the rest by physically elevating your body to a higher position. Think of thrones, some of the most elaborate chairs still in existence. Such associations of chairs with stature found their way into the 20th-century office, where the boss's chair tended to be more grandiose and better designed than the rest of the workers'. It was when the ergonomic office chair as we know it today appeared that such rigid office hierarchies started to crumble.

"Sitting on a chair was a declaration of power and higher status: you were setting yourself apart from the rest by physically elevating your body to a higher position."

One of the most recognisable office chairs was manufactured by American furniture company Herman Miller (officially MillerKnoll Inc.), which has been at the forefront of office design innovation since the 1960s. Having produced iconic mid-century designs like Charles and Ray Eames's chairs and Isamu Noguchi's tables, the company shifted its focus to office furniture in the '60s, with the success of Robert Propst and George Nelson's "Action Office". This was a hybrid of a closed and an open-plan office, in which each worker had a few dedicated

The Mirra chair, designed in 2003, is made with recycled materials and the use of green power.

high-end furniture-workstations, including the innovation of the standing desk, to encourage movement and communication. This dynamic office design didn't catch on and its subsequent adaptations led to what became known as the "cubicle": small, enclosed desk spaces divided by fabric partitions, these dominated the office landscape until the late '90s. In 1994, Herman Miller introduced the Aeron chair from designers Bill Stumpf and Don Chadwick, which incorporated innovations such as the breathable and flexible mesh material Pellicle and a tilt mechanism where seat and backrest could be adjusted simultaneously with one motion for greater support when reclining. Such features were based on decades of scientific research with a focus on older adults' needs, and in collaboration with specialists in orthopaedic and vascular medicine. The Aeron marked a turning point for office chairs, setting the standards of ergonomic design. Coinciding with the coming of the internet and start-ups, it became a bestseller and a Silicon Valley status symbol.

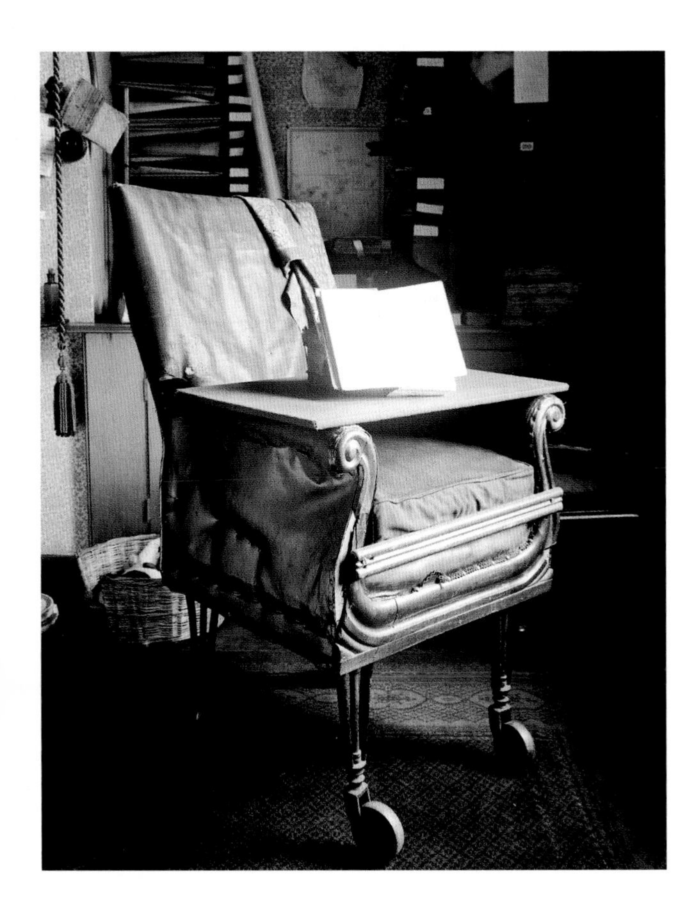

Charles Darwin's wheeled armchair.

Aeron's design was influenced by the concept of Universal Design. Its inclusive and accessible principles are credited to architect Ron Mace, who explained it as "the design of products and environments to be usable by all people, to the greatest extent possible, without the need for adaptation or specialised design". When applied, features such as step-free access, wider entrances and contrasting colours and fonts are introduced in private and public spaces, assisting with the elimination of barriers in the built environment that Mace experienced first-hand as a disabled person in the '80s.

Ergonomics – from the Greek words *ergon* ("work") and *nomos* ("natural laws") – is the science of designing a workplace for the body's natural needs. The term and its definition first appeared in 1857 in Polish scholar Wojciech Jastrzębowski's publication *An Outline of Ergonomics or the Science of Work Based Upon Truths Drawn from the Science of Nature*. But connecting work with health has an older history than that. Italian physician Bernardino Ramazzini wrote about occupational health in his 1700 book on workers' diseases. During the 20th century's two world wars, ergonomics research was applied in improving working conditions and efficiency in the military. Research into accidents in the US Air Force showed that these were mainly due to bad design of the altimeter's interface, which incapacitated the pilots' abilities. In the UK, legislation has been in place since the 1970s describing seating as work equipment, thus placing obligations on employers, seating manufacturers and suppliers to support the health and safety of people at work by preventing hazards due to unsuitable seating such as discomfort, back pain and upper limb disorders.

If chairs are "echoes of the body", as Galen Cranz stated in her 1998 book *Chair: Rethinking Body Culture and Design*, ergonomic, medical and design research has been disproportionately conducted by and for male bodies, creating products unfit for the specific needs of more than half of the global workforce, and putting women at higher risk of occupational injuries. In an effort to address this imbalance, the International Ergonomics Association created the Gender and Work Technical Committee in 2006, with a focus on integrating sex and gender into ergonomics research and work environment design.

American artist and designer Ray Eames was not the only woman to design an iconic chair; Florence Knoll's armchairs and Charlotte Perriand's tubular steel chairs are now in museum collections around the world. However, many other women designers remain overlooked by a male-dominated design history, the legacy of which lives on in the current gender gap in furniture design. In 2019, British designer Laila Laurel humorously addressed the gendered nature of sitting with her award-winning "anti-manspreading" chair, crafted so that men sit with their legs closed.

"A global pandemic in 2020 left its mark . . . The closure and downsizing of big companies' offices fuelled a booming trade of stranded office chairs."

The green Mirra chair (seen on page 76) from 2003, designed by Herman Miller's German firm Studio 7.5, was the first office chair to be developed using "cradle-to-cradle" design. Conceived by chemist Michael Braungart and architect William McDonough, "cradle-to-cradle" is a circular economy concept combining design and science to eliminate waste. This particular chair is made from up to 96 per cent recyclable steel, plastic, foam and textile, and 42 per cent recycled materials, while its manufacturing utilised 100 per cent green power from wind turbines and captured and treated gases emitted from organic decomposition in landfills.

Some 20 years after the chair's conception, most furniture still ends up in landfill, proof of a "fast furniture" mentality. As the climate crisis deepens, sustainability is higher on the agenda of scientists, designers and more consumers. The use of eco-friendly materials such as various organically grown vegetable fibres, like wicker, bamboo and rattan, and the repurposing of plastic waste (otherwise harmful to the environment) are becoming more popular while second-hand furniture markets and "upcycling" (the repurposing of unwanted products) flourish.

For the past 100 years, office chairs have developed alongside our relationship to work and design, reflecting changing social, scientific and cultural conditions. A global pandemic in 2020 left its mark on office working culture and design. The closure and downsizing of big companies' offices fuelled a booming trade of stranded office chairs. For example, Twitter sold chairs for thousands of dollars along with other furniture from its San Francisco headquarters in a public auction in early 2023. Iconic office chairs like the Aeron, once symbols of the tech economic boom, ended up stacked in warehouses, symbols of a post-pandemic economic recession.

Office chairs entered our houses, bringing the domestic sphere and the workplace closer than ever just as we spent more time "sitting" in the digital world. This blurring of boundaries between physical and virtual working spaces led to office and chair design innovation characterised by flexibility, adaptability and customisation with an increasing focus on wellbeing and safety.

On top of investing in digital tools, companies are now looking at agile and multipurpose office spaces and furniture (e.g., spaces for collaboration coexisting with spaces for quiet work and privacy) to attract professionals back to the physical workplace, echoing ideas of the Action Office and ergonomic principles of health. Turning to technology could also assist with offsetting the hazards of sedentary work life. Smart office chairs equipped with AI-powered sensors that detect our posture and warn us to improve it are in development. They may be an aspect of life overlooked by many, but it's interesting to consider what the office chairs of the future will say about us and our world.

The Tokyo "Isu-one Grand Prix" – a two-hour endurance race on office chairs.

PERSONAL COMPUTER

Computers can be found throughout our day-to-day lives, not just on our desks and in our pockets, but embedded into every conceivable industry. Their success is the result of decades of fast-paced innovation, the miniaturisation of components, hobbyists' passions and user-focused design. Introduced in 1981, the IBM Model 5150 opposite was the catalyst for the personal computer (PC) industry we know today. At the time, IBM was the world's largest mainframe computer manufacturer. Despite their late entry into the PC market, this machine became one of the most popular computer design standards in the world. Its success, selling around 200,000 machines in the first year, marked the end of several personal computer companies that could not compete with IBM's market dominance.

When we talk about computers today, we usually mean electronic devices that can be programmed to carry out many different tasks; in other words, universal stored-program computers. The first were built in the 1940s, but there were "computers" before then. The word *computer* was used from the 17th century to describe someone employed to perform the repetitive calculations needed for tasks such as producing astronomical tables. By the 19th century, computers were still commonly working for observatories and had spread to other areas, such as tracking weather patterns and businesses handling large volumes of data. Often forgotten, many of them were women, their work unacknowledged or uncredited as it was considered low-skilled.

"The word *computer* was used from the 17th century to describe someone employed to perform the repetitive calculations needed for tasks such as producing astronomical tables."

Before electronic computers, computation was carried out laboriously by hand, often assisted by printed mathematical tables and mechanical calculating machines. In other cases, one-off devices computed specific problems. But unlike today's computers, these single-purpose machines could not be reprogrammed to handle different problems. Then came the mathematician Alan Turing. In 1936, Turing wrote a seminal mathematics paper, "On Computable

Around 200,000 units of the IBM Model 5150 were sold in the first year of its release.

Numbers, with an Application to the Entscheidungsproblem", which came to be seen as the theoretical basis for today's computers. In it, Turing imagined a single machine that could compute any problem, effectively uniting all human or problem-specific "computers" into one universal device. A few years later, these modern computers started to be built to meet wartime needs. Built for their number-crunching power, they were used to break codes and design new weapons.

While much of this work remained secret until the 1970s, engineers and scientists who had been mobilised for the war effort returned to university laboratories and government departments inspired by the potential for electronic computing and armed with technical know-how. By the 1950s, this enthusiasm had spawned several bespoke and commercial mainframe computers in Britain and the US, which were being used for their data-processing abilities in sizeable institutions, companies and universities. These large machines, often the size of a room, were expensive to run and had to be shared among several people to be economically viable. While computing power was limited to those who could afford it, each machine made during this period featured innovations in storage, speed and applications.

Users would not interact directly with the machines, but would prepare tasks for the computer on other equipment, such as punched cards, which would then be input into the computer, often by a trained attendant. Despite being significant scientific tools, these early mainframe computers were slow, sometimes taking hours or days to

Marlyn Wescoff (left) and Ruth Lichterman operating ENIAC, one of the first general-purpose electronic digital computers, 1946.

process one problem. The slow speeds were partly overcome by the introduction of time-sharing in the mid-1960s, which let several people use one mainframe computer simultaneously through their own computer terminals, and encouraged the idea of making smaller machines for single users.

Early mainframe computers consisted of valves or transistors, which were wired together on circuit boards. In the late 1950s, electronic manufacturers developed ways of creating whole circuits on silicon chips. These integrated circuits, or "chips", were smaller, more reliable, less energy-intensive and cheaper to make than older electronic components. In 1971, Intel released the first commercially produced "microprocessor" – the 4004 – a complete computer processor on an individual chip. This development proved to be the gateway to the popularisation of cheap, easy-to-use computers.

One of the first "microcomputers" to use these new processors was the Altair 8800, released in 1974 by a small company called Micro Instrumentation Telemetry System (MITS). Unlike today's personal computers, the buyer had to build the machine from a kit, and each was quite expensive at $379. It did not have a screen and was programmed in binary code by flicking switches on the front panel. The output was then shown as flashing lights. Despite its rudimentary design, the company was overwhelmed with orders and struggled to keep up with demand, chiefly from amateur computer hobbyists. The Altair 8800 inspired the entrepreneurs behind some of the world's biggest computing companies, including Bill Gates and Paul Allen, who founded Microsoft, and Steve Wozniak, the co-founder of Apple Computer, Inc.

The personal computer industry truly began in 1977, with the release of three mass-produced personal computers: Apple II, by Apple Computer, Inc.; the Tandy Radio Shack TRS-80; and the Commodore Business Machine Personal Electronic Transactor (PET). While initially aimed at hobbyists, these machines created a whole new market for computers. They were much cheaper than mainframe equivalents because they had smaller microprocessors and limited memory. Part of their appeal was that they came preassembled with a keyboard, screen, a means of programme storage, and BASIC programming language pre-loaded.

Seeing the potential for the PC market to grow from a hobbyist's pastime to a billion-dollar industry, IBM created the 5150 in record speed using an open architecture that allowed them to buy components from other companies, including the Intel 8088 microprocessor and MS-DOS, Microsoft's operating system. This design approach encouraged other manufacturers to produce clone machines that ran the same software, standardising computers in the process and securing Intel and Microsoft's dominance in the PC market.

The popularity of personal computers relied on user-friendly software that helped people with real-life tasks. The main markets were games, education and business, where the ability to manipulate large volumes of data quickly and efficiently had obvious benefits. A significant development was the graphical user interface of GUI (pronounced "gooey"). First proposed in 1968 by Douglas Engelbart and commercialised in the Apple Lisa in 1983, the GUI replaced the need to type instructions into a computer with windows, icons and pull-down menus operated using a mouse. Still the industry standard, GUIs made computers easier to use and accessible to wider society, without needing to know the inner workings of the machine.

Despite a wide range of software applications, computers in the mid-1980s were isolated machines. This changed with the introduction of CD-ROMs, which provided access to encyclopaedias and multimedia entertainment, and computer networks, which let users contact one another through emails and chatrooms. The internet and World Wide Web recast the personal computer as a machine of the information age, and it continues to profoundly shape our lives.

INVISIBLE HEARING AID

During World War I, a woman named Mary was stopped and arrested while taking a holiday stroll down the Atlantic City Boardwalk. Her arrest wasn't because she was acting suspiciously, or disturbing the peace, but because of her hearing aid: a black box weighing about 3 kg (7 lb) which authorities assumed was a wireless spying device.

Today, hearing aids like those pictured opposite are tiny, almost invisible devices that hide on or within the ear, and not the huge, conspicuous boxes of the past. The developments that led to this transformation not only changed the lives of hearing aid users, but also changed the world of miniature technology more broadly.

For electronic devices to be small enough to carry or wear, they need to have components that are functional but miniature, circuits that fit closely together, and batteries that can power the circuits without taking up too much space. Hearing aids served as the test bed for much of the innovation required to shrink down electronics, and were one of the earliest uses of many of the fundamental components of miniature technology. This progress was driven not only by the hearing aid industry itself, but also by the users who were actively involved throughout, as inventors, campaigners and manufacturers.

"Hearing aids served as the test bed for much of the innovation required to shrink down electronics, and were one of the earliest uses of many of the fundamental components of miniature technology."

Hearing aids are devices that amplify sound and distinguish speech from background noise. They are incredibly useful for many deaf people who have specific types of acquired hearing impairment. There are types of hearing impairment for which hearing aids will not work, while members of the Deaf community – a cultural group who use sign language as their primary language – mostly choose not to use hearing aids. However, the use of hearing aids today is widespread, and many deaf people throughout history have had a strong desire for practical, comfortable and functional ways to improve sound amplification.

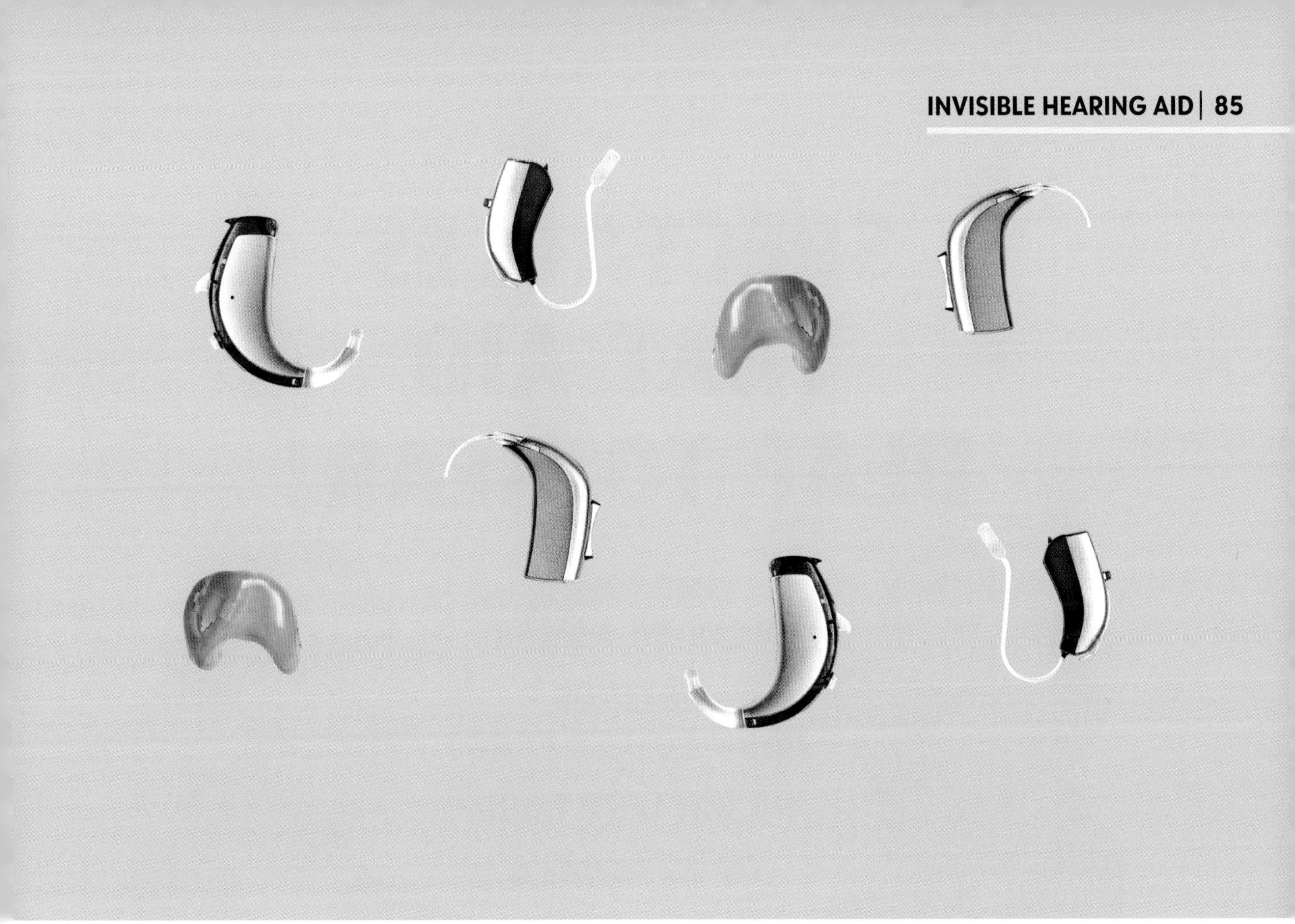

A collection of various modern hearing aid devices.

Early hearing aids took the form of horns or trumpets, which caught sound waves and transmitted them to the ear. The world of electronic hearing aids, however, begins at the end of the 1800s with the invention of the telephone. After seeing Alexander Graham Bell's telephone, Thomas Edison, who had a hearing impairment himself, claimed that he was unable to hear through it properly, and was motivated to create a better component. In 1898, Edison became one of the first inventors of the carbon transmitter, a metal and carbon disc that converts sound waves into electrical signals. The carbon transmitter went on to find a home both in telephones and in the first electronic hearing aids.

This revolution in sound amplification was only the beginning. Hearing aids in the late 19th century were large and obvious, and the use of them largely stigmatised. Many users spoke of how it advertised their hearing impairment to the world, leading to intrusive questions or embarrassment among friends. Frances Warfield, a member of the New York League for the Hard of Hearing (and friend of Mary, the Atlantic City holidaymaker mistaken for a spy), pressured AT&T to make electronic audiometers, but still expressed concern at looking like a "walking telephone". With this strong demand for hearing aids to be as small as possible, and taking into account the large potential customer base and relatively simple

SPECTACLES THAT AID DEAF TO HEAR!

IMPROVED HEARING POWER!

NOTHING IN OR ON THE EARS!

NO BATTERY CORD!

NO HEADBAND!

The new Fortiphone 'SPECTA-PHONE' Bone-conduction Hearing Aid improves the power and clarity of bone-conduction hearing and makes a headband unnecessary. It is therefore more comfortable and convenient. Recommended even to the hard-of-hearing whose sight needs no assistance.

TEST IT FREE!—Call, Phone, Write, or Post this

Only by personal test can you judge the clarity, comfort, and convenience of this and other new Fortiphone Aids to Hearing. Accept our offer of a Free Consultation and Test at our Consultation Rooms (*hours* 9-5.30, *Sats.* 9-12.30) or in your own home. Please call, telephone REGent 2024, write, or post this coupon today.

Full year's Guarantee! Apply within 30 days for Money-saving Offer!

'SPECTA-PHONE' is a trade mark

—FREE HOME TEST COUPON—

to FORTIPHONE LIMITED (Dept 379)
Fortiphone House
247 REGENT Street, London, W.1

I would like to know more about your new 'SPECTA-PHONE' Hearing Aid and Free Home Test Offer advertised in The Queen, 3/6/53.

Name ....
(*Mr, Mrs, Miss, or title*)

Address ....

....

Phone: REGent 2024 **BLOCK LETTERS PLEASE**

A 1950s advert for hearing aids disguised as spectacles.

circuitry, hearing aids became the perfect site for miniature technology innovation.

Carbon transmitter hearing aids did shrink over time, though they were not the most efficient at amplifying sounds. A step forward in function came from the use of vacuum tubes to replace the carbon transmitter, but this led to an initial step backwards in terms of size. Vacuum tubes are sealed glass tubes that create a vacuum to allow the free passage of electric current, and these early tubes were large and required similarly sizeable batteries. The first vacuum tube hearing aid in 1920 weighed roughly 3 kg (6½ lb), and in 1923 they still needed to be carried around like a small suitcase. As the vacuum tube aids started to get smaller, users got creative with the ways they disguised them, hiding the microphones in their hair or sewing the battery packs into their skirts. But it wasn't until 1938 that new subminiature vacuum tubes were developed and released by American electronics engineer Norman Krim, at the electronics and technology company Raytheon.

By the 1930s and '40s, hearing aids with smaller tubes and batteries could be worn across the body, in multiple connected parts. They were still cumbersome and often undesirable, however, despite manufacturers' attempts to encourage sales by producing adverts with glamorous women strapping on their hearing aids over silky nightgowns.

In 1936 Paul Eisler, an Austrian-Jewish refugee and engineer residing in London, invented the printed circuit. A revolution in miniature electronics, printed circuits allow all components to be printed and mounted on a single board, producing a smaller and sturdier circuit without the need for wires. Initially used for military purposes, the invention was released in 1948 and found its first commercial application in the "Solo-pak" hearing aid from Allen Howe Electronics. With smaller circuits and tiny new "button batteries", the Solo-pak marketed itself as "small as a cigarette packet".

"Hearing aids in the late 19th century were large and obvious, and the use of them largely stigmatised."

Around the same time, the transistor was invented, a revolutionary component for controlling electronic flow with the potential to replace vacuum tubes. Norman Krim, the previous developer of the subminiature vacuum tube, saw the possibilities for transistors to transform hearing aids. The earliest transistors were not appropriate for this purpose, so Krim set his Raytheon engineers to develop a better, more stable version. By 1951, the junction transmitter was ready to be manufactured and sold to a large number of hearing aid companies. George Freedman, who managed the development and manufacture, decided to exclusively employ hearing aid users to manufacture the transistors. Hearing aids had previously made use of a mix of transistors and vacuum tubes, but by 1953 the world had its first all-transistor device.

These new and smaller hearing aids were much more practical to wear than their predecessors, but it is clear that the stigma attached to hearing aid use was still present, as many new aids were designed to be disguised in creative ways, on clothing or as

Vintage *Life* Magazine 11 January 1943 Issue Advert, USA.

jewellery. Throughout the 1940s and '50s, hearing aids for men could be hidden inside suit pockets, or disguised as an insignia tie clip or a fountain pen. Women could hide their hearing aids on a hair clip or inside a stylish hat or scarf, or buy aids that looked like a brooch or pearl necklace. By 1959, almost half of all hearing aid sales in the US were for an aid that sat on a pair of eyeglasses, suggesting that hearing aid users were more comfortable wearing glasses like those who required visual correction than advertising their deafness with other devices.

In 1958, the Integrated Circuit was invented, also known as a microchip. With the entire circuit and all accompanying parts formed of one small piece of the same material, it removed the need for separate transistors and other components. This allowed for even smaller electronics, and found its first commercial home in the hearing aid. By 1964, the first behind-the-ear aid using an integrated circuit became available, and the size and practicality of hearing aids continued to improve from there.

Hearing aids were the first commercial application of the transistor, printed circuit and integrated circuit, and were an early use of vacuum tubes and miniature batteries. In 1971, the place of hearing aids at the forefront of miniature technology came to an end with the invention of the microprocessor, a chip that could perform complicated computing functions within its tiny circuits. As the fields of electronics and computing merged, the greatest innovations began to take place in the world of computers instead. Hearing aids finally adopted microprocessors in the late 1980s, and continue to improve today, but their place as the testing ground for new technologies has come to an end.

The technological innovations and knowledge born out of the race for smaller hearing aids

"Throughout the 1940s and '50s, hearing aids for men could be hidden inside suit pockets, or disguised as an insignia tie clip or a fountain pen."

were linked to a desire for more comfortable and practical devices, but also undeniably linked to the stigma of hearing aid use. Throughout the 20th century, users often wanted hearing aids that were disguised as more socially acceptable wearables or hidden inside clothing. In fact, the initial stigma was large enough that early examples of miniature technology outside the hearing aid field struggled to take hold. The first pocket radio didn't prove hugely popular, in part due to the fact that people thought it looked too much like a hearing aid. Are the transparent, invisible, in-ear aids we have today a revolution in comfort and practicality for users, or are they a sign that there is still a pressure to hide what might mark you out as "different"? Perhaps they are both.

What cannot be understated, however, is the importance of the hearing aid inventors, manufacturers and users and their work. Their legacy is the miniature technology we rely on today, such as our phones and laptops, and many different types of wearable devices. Those early steps into a miniature electronic world, and the problem-solving required to make small technology that can be carried around with us, are intricately linked to the race for ever smaller hearing aids.

The components of modern miniature hearing aids, which are nearly invisible when worn.

MICROWAVE OVEN

Despite being a common fixture of many kitchen countertops, the microwave oven and microwave technology have a far less mundane history. A clue to their origins is found in the Dutch word for microwave oven: when you reheat your cold coffee in the Netherlands, you put it in a *magnetron*.

Microwaves are a type of electromagnetic radiation with wavelengths between approximately 30 and 0.03 cm, though despite their name they are found at the lower frequency end of the electromagnetic (EM) spectrum. James Clerk Maxwell's 1873 theory of electromagnetism was first demonstrated by Heinrich Hertz, who generated radio waves in 1888. A few years later, Bengali polymath Jagadish Chandra Bose became the first to generate and experiment with microwave frequencies, though interestingly, the first recorded use of the term *microwave* wasn't until 1931.

The turn of the 20th century was an era in which many new sources of energy were being developed, including EM radiation. No one knew the potential of these invisible waves yet, and the thought of their possible uses in technology intrigued people. H.G. Wells wrote of a Martian heat ray in his 1898 novel *The War of the Worlds*, and from then on the concept of "rays" (death rays, ray guns, etc.) became firmly embedded in the science fiction genre.

"Mainstream media after World War I was filled with stories of the kinds of weapons that could be expected in future conflict."

The idea of a "death ray" was somewhat ambiguous, with no singular concept describing what would power them or what they could do – for instance, would they kill or merely disable an enemy? In a futuristic genre like science fiction, writers could take liberties with a theoretical death ray's capability.

Almost as soon as World War I ended in 1918, people began talking of "the next war". Mainstream media was filled with stories of the kinds of weapons that could be

The domestic microwave developed from military use of radar technology.

H2S Mark 9A Yellow Aster Bombing Radar – this scanner was fitted to an Avro Vulcan V2 bomber plane.

expected in future conflict, including speculation of lethal rays, perhaps powered by electricity to generate heat with fatal impact. Scientists tried hard to debunk claims of death rays to no avail, despite multiple conclusions that generating enough energy to create a ray effective as a military weapon was technologically impossible.

Among this sensationalism was speculation about the new technology Germany could be developing, despite the restrictions imposed on them by the Treaty of Versailles. In 1923, a few instances of French commercial airplanes experiencing sudden engine failure over a certain region of Bavaria added more fuel to the fire. Rumours rapidly spread that the Germans had developed some kind of ray that would short-circuit a gasoline motor.

Wary of falling behind in the technological arms race, in 1935 the British Air Ministry put forward a prize of £1,000 (upwards of £50,000 in today's money) to anyone who could build a "death ray" that could kill a sheep from a distance of 100 yards. Although many inventors attempted to create death rays with varying success, this prize money was ultimately never collected.

Around this time, Britain was attempting to turn the scientific concepts of EM radiation into real technology. Radio researcher Robert Watson-Watt had been busy studying the feasibility of "radio destruction", or the idea that high-energy EM radiation could destroy an aircraft or put its crew out of action, on behalf of the British Air Ministry. By 1935 he had determined that it wasn't possible, but along with his colleague Arnold Wilkins realised that EM waves reflecting off aircraft could be used to determine their location. This formed the starting point for microwave radar technology in Britain.

No one country led the development of radar, but the threat of another war in continental Europe incentivised many countries to continue working on technology that could be utilised in warfare, especially as World War I had shown how important aircraft would become in future conflict. The term "RADAR" – an acronym for "radio detection and ranging" – wasn't actually used until the US Navy introduced it in 1940.

During World War II, radar technology was vital, and the British used it in a ground-based, early-warning radar network – the first in the world – codenamed "Chain Home". The system detected and tracked aircraft, giving advance warning of the German Luftwaffe approaching from France.

The invention of the cavity magnetron in 1940 by British physicists John Randall and Harry Boot marked a huge leap forward in microwave technology. Cavity magnetrons were the first practical device that could produce short waves (microwaves) of enough power to run miniaturised radar systems. Randall and Boot's first working example produced hundreds of

Original cavity magnetron, developed by John Randall and Henry Boot in 1940.

From the 1970s, the microwave oven became a popular appliance as increasing numbers of women went into the workplace.

watts of power at just a 10 cm (4 in.) wavelength, meaning that the radar systems could be made smaller than ever before.

This put Britain at the forefront of microwave radar, and the technology was a closely guarded secret during wartime. The H2S radar system was the first airborne, ground-scanning radar system used to identify targets for bombing. Although its larger range was a big improvement on previous radio navigation aids, the British initially worried that using it was too risky in case the Germans got their hands on the new cavity magnetron technology. Pilots flying aircraft fitted with H2S were told to destroy the system if they thought they were at risk of crashing or being captured in occupied countries – a hard ask, as they were mounted on the outside of the aircraft. When the technology was offered to the US in return for financial and industrial support to manufacture it, the magnetron travelled across the Atlantic Ocean in a black box that was designed to sink in the event that the ship carrying it came under attack.

Even with all these precautions, it didn't take too long for the magnetron technology to make it into the Germans' hands by way of a downed bomber in 1943, though several countries including Germany had been independently developing their own radar and magnetron technology throughout the 1930s anyway.

The first practical system of microwave ovens came during post-war manufacturing, when US industrialised production made the change from military to commerce. Percy Spencer, working on an active radar set for American defence contractors Raytheon, discovered that a chocolate bar carried in his pocket had been melted by the microwaves and realised that it was possible to cook efficiently using these waves – a process called "diathermy". Microwaves cause the water molecules in food to vibrate, heating it up.

Suddenly there was a whole new domestic use for these short-length waves. But it took a while for microwave ovens to reach our countertops. Raytheon produced the first commercial microwave ovens in 1946, intended for use in restaurants or for reheating meals on aircraft. Called "Radaranges", these were large, expensive appliances that had to be continuously water-cooled. Japanese company Sharp developed their first microwave oven, the R-10, in 1961, and became the first company to mass-produce microwaves a year later, though they focused on commercial use. It wasn't until 1967 that the Amana Radarange countertop model was introduced as the first compact microwave oven marketed for home use, though it remains larger than the standard size we have today. It wasn't cheap either – around $495 at the time, which is now over $4,000.

Microwave ovens increased in popularity in more developed countries throughout the 1970s and '80s, with around one in four US homes owning one by 1986. The increase of microwave oven ownership was linked to social changes such as the increase of women in the workplace. The task of preparing meals in households still commonly fell to women, and microwave ovens could be utilised to save them time, as they could reheat bulk-cooked food or use frozen or pre-packaged meals. In 1988, nearly 12 per cent of the new food products released into the US market were specifically designed for microwaving. Additionally, dual-earning households were more likely to have the disposable income needed to purchase a new appliance, as they were still quite pricey.

So, is there a future for this technology, or have we peaked with preparing popcorn for at-home movie nights?

In the 1950s, scientists used diathermic heating via microwaves to gently "reheat" small mammals such as rats and hamsters after they had been frozen to temperatures of below 0°C (32°F). This was part of research into cryo-biology, and they managed to successfully reanimate many of the test subjects, some several times over.

Cryopreservation is still an important field of research in medicine today, as it could allow us to preserve tissue or blood for organ donations or blood transfusions. However, scientists have found more effective ways of reheating tissue, since diathermy can produce hotspots, so it's unlikely that operating theatres will be kitted out with microwave ovens any time soon.

TINNED FOOD

Imagine you find yourself lost in some incredibly remote place, perhaps atop a mountain among an endless range of peaks, or on a far-flung island in a stormy ocean with only the sea, sand and sky for company. Wherever you envisage, the fact remains the same: you are alone, and it will be some days or even weeks before help finds you.

While ambling about trying to remember if you ever learned how to start a fire with pieces of stick or rock, you come across something so valuable, and so utterly convenient, that you briefly consider falling to your knees with joy.

It is a stash of tinned food. A neat pile of tin cans, each one a closed cylinder of silvery metal – small enough to fit in your hand, large enough to contain a substantial amount of nourishment. The packaging illustrates what is safely preserved within. Each (hopefully) has a ring pull on the lid, which can be peeled back with a little force to reveal the beans, tuna, soup, or any other food you might desire. It might even contain a sugary fizzy drink.

The tin can is an example of a technology that is slightly mundane when unceremoniously stacked in the kitchen cupboards of industrialised societies. Yet in this remote landscape, it is one that appears like a cherished time capsule.

The tin can in the kitchen cupboard does the same thing as the tin can found on the side of an isolated mountain peak: it prolongs the usefulness of food, keeping it edible for far longer than microorganisms and spoiling would normally allow. It provides an escape from the constraint to eat only what can be sourced, harvested, caught or prepared there and then. It can be transported, bought, sold or exchanged with ease. It neatly packages up a store of energy that can be accessed and eaten no matter the season. And once made and sealed, it requires no extra energy to keep it unspoiled.

"Once made and sealed, the tin can requires no extra energy to keep it unspoiled."

Prolonging the shelf life of food is important for achieving healthy levels of nutrition, minimising food waste and improving food security. It is doubly important in times of scarcity or upset, when the complex food systems on which so many rely can become disrupted.

Tin can from family war supplies, 1940–45.

The tin of rolled oats (opposite) was manufactured in the early 1940s by the Quaker Oats Company. It formed part of the wartime emergency food stocks of the family who donated it to the museum, and contains around 14 servings that could have been prepared simply with a little water or milk. Cooking suggestions in eight languages are printed on the label.

Tinned food is an early example of modern industrial food preservation, and tin cans have wound their way into both daily life and popular culture, from the well-trodden genre of post-apocalyptic fiction to cartoons and art galleries. Emily St. John Mandel's 2014 novel *Station Eleven* features a man receiving a tip-off about a rapidly spreading swine flu pandemic; he spends a fraught 40 minutes racing to fill trolley after trolley with "cans and cans of food… anything that looked like it might last a while." Andy Warhol's series of paintings *Campbell's Soup Cans* (1961–62) became synonymous with the Pop Art movement, and accomplished the feat of transforming the tin can from everyday functionality to high-profile modern icon, accorded centre stage. But how did this 200-year-old industry first come about?

Before tinned food, millennia-old food processing techniques included salting, drying, smoking or burying in ice. The ancient Egyptians, who took their skills to another level with embalming, the Greeks and Cretans, who fired *pithoi* to store and transport oil and grain, the clarified butter of ancient India and north and east Africa, and the bog butter of Ireland and Scotland all provide incontrovertible evidence of long-established practices of food preservation. Pickling and fermenting were also commonplace before industrialisation.

In Europe in the early 1800s, the link between microorganisms and food spoilage was not yet known (and would not be confirmed until Louis Pasteur's experiments in the 1860s). Those who went on long journeys at sea had to be content with a diet that mostly consisted of hardtack – biscuits made from flour and water, which were so dense that they could break your teeth if they weren't soaked in water first. The British, Dutch and French navies were all keen to flex their military might in the name of colonialism and national supremacy, and this required a solution to the difficulty of feeding thousands of troops far away from their home country's food supplies.

Early tin can made by Donkin, 1812.

In 1810, Frenchman Nicolas Appert published his findings on a new method of effectively sterilising food by heating it in sealed glass bottles. Later the same year, Peter Durand modified this approach, replacing glass with tin, and the patent found its way to an engineer called Bryan Donkin. With his collaborators, Donkin set up a canning factory in Bermondsey, South London, in 1812. Here, a small number of people each made six tin cans an hour, and the fledgling industry came into being. The holey, rusted tin can on page 97 is the oldest known survivor from that period and would have been filled with around 3 kg (6½ lbs) of veal or beef. Donkin's tins were sold primarily to the military, ending a reliance on unsatisfying hardtack for many. (Having a bayonet to hand to carve into the sealed tin was an optional bonus; dedicated tin openers did not yet exist, and early designs would not be patented until the 1850s. A hammer and chisel was the recommended approach.)

The manufacturing methods used then are largely the same as today. The tin can was made and filled with food, then sealed and placed in water, which was brought to the boil. The lid was taken off very slightly to let out hot air, and then the whole thing was sealed again, often soldered shut with lead. (This toxic metal sometimes managed to leak into the food itself, as is purported to have happened on John Franklin's ill-fated

"Early designs for tin openers would not be patented until the 1850s. A hammer and chisel was the recommended approach."

Space food made by Heston Blumenthal for Tim Peake, 2015.

Arctic expedition, which disappeared in 1847, with tins supplied by Stephan Goldner, a competitor of Donkin's.) Finally, the tins were kept at 90–100°C for up to a month, ensuring that no microbe had a chance of survival. While Donkin's tins involved commendable levels of quality control, Goldner's did not, and at the time the much-publicised Arctic disaster had a real impact on public confidence in and adoption of tinned foods, with the introduction of domestic refrigerators also beginning to vie for attention.

No longer marrying the promise of a nutritious meal with the somewhat less inviting threat of lead poisoning, the production of modern-day tinned food is thoroughly well-managed. And a couple of centuries on from its inception, tinned food is now a staple for those who go on expeditions of a different kind. Chef Heston Blumenthal cooked up tins of space food for astronaut Tim Peake in 2015 (see above). As in more terrestrial settings, astronauts aboard the International Space Station like to come together to eat.

As space missions become more ambitious and far-reaching, there will be greater pressure to provide longer-lasting food. When Mars is closest to Earth, it remains a staggering 55 million km (34 million miles) away. If a crew wishes to head there, they will need enough food to last them at least three years. Presently, tinned foods work small miracles in this arena: high-acid foods such as tomatoes and fruits keep well for 12–18 months; those low in acid, like vegetables or meats, stay as good as new for 2–5 years. And many people have anecdotes of various tins lasting far longer.

The desire of Elon Musk (and others like him) to colonise Mars will require longer-lasting foods.

Transcending its military origins, tinned food has an important place in our future. Culture and food are intimately linked, and tin cans can be used to sustainably transport foods that are about more than just nourishment, but that also have value in their strong associations with the idea of home and with their native landscapes, even when eaten far from their original source. Wherever people go, wherever they migrate, there will be gaps in the available foodstuffs. The preservation and easy transportation of foods that also provide a slice of human culture can become of heightened significance in a world where the ongoing effects of climate change are intertwined with enduring legacies of colonialism, and where the lands and livelihoods of those living in warmer climes are being disproportionately affected.

Fruit in tin cans last up to 18 months and meat for as much as 5 years.

Back on your mountain or island, the search party finds you happy and well. You have eaten your way through a series of foods that, given the circumstances, even the harshest restaurant critic could not turn their nose up at. Not only is it a monumental relief to be found, but it is also a moment – before replenishing the stash for the next lost person who winds up there – to open the remaining tins with your rescuers and eat among company once more. The best meals are shared. And tinned food has saved or enlivened many a meal the world over and beyond.

LIGHTBULB

For many of us, light is accessible at the flick of a switch. But for much of history – and for many people living across the world today – it took a lot more work to see in the dark. Candles, oil lamps and fires were the main sources of light, limiting you to very small areas of illumination once the sun went down (unless you could afford an endless supply of candles!). The majority of the 20th century was dominated by the light of incandescent bulbs, which work by heating a filament with electricity until it glows. The filament is encased in a glass bulb that keeps out oxygen to prevent it from burning out.

The bulb opposite is one of the first practical incandescent lightbulbs, emitting the equivalent light of 13 candles over the course of 1,390 hours (just over 57 days). It was patented by the American inventor Thomas Edison in 1879 and helped to put his name in the history books as one of the most credited inventors. However, it took almost the entirety of the 19th century to perfect this idea, with numerous inventors contributing designs, each building on their competitors' ideas in a battle to produce a practical light source – a "battle of the bulbs", you could say. Edison's patent was awarded as an improvement of electric lamps, not an invention – a subtle differentiation that acknowledged all the existing work his success was built upon.

This battle of the bulbs began at the start of the 1800s when Sir Humphry Davy – an English chemist – demonstrated that a small burst of light could be produced by a platinum strip when electrified. Little did he know the potential of this discovery, which led to one of the main components of the lightbulb: the filament. Davy moved onto other, less successful, ways of producing light, but many inventors took his filament and tinkered with it to try to create a more successful design. The first lightbulb patent was awarded in 1841, almost 40 years before Edison's patented incandescent bulb brought light to the masses.

"It took almost the entirety of the 19th century to perfect the lightbulb, with numerous inventors contributing designs, each building on their competitors' ideas in a battle to produce a practical light source."

The battle didn't end with Edison's 1879 lightbulb, however. His biggest rival was the English inventor Sir Joseph Swan, who had come up with a very similar carbon filament design around the same time. The two men initially formed competing companies and spent years in their own mini battle to achieve "world firsts". Swan's home in Gateshead,

A very early Edison carbon filament lamp, 1879.

After inventing versions of the lightbulb separately, Thomas Edison and Sir Joseph Swan merged their companies in 1883.

UK, was the first private residence to have electric lightbulbs installed, and his bulbs lit the first street in the world with electricity: Mosley Street, Newcastle, in February 1879. Edison achieved a similar feat later that year, lighting up New York's Christie Street on 31 December.

At the start of 1881, Swan also became the first person to commercially manufacture incandescent lightbulbs, going on to produce over 1,000 lightbulbs for the Savoy Theatre in London (the first public building to be entirely lit with electric lighting when it opened in October 1881). As usual, Edison wasn't far behind with New York's Pearl Street station. Opened in September 1882, it was the first commercial power plant in the US and provided electric light to multiple buildings within a 50 by 100 feet area (approximately 15 by 30 m). While Swan was ahead of Edison in implementing the infrastructure of his designs, Edison was more concerned with legalities and ensuring he held all the rights. The men eventually came to blows in the courts over the similarity of their patented bulbs, settling out of court by merging their companies to form the Edison and Swan Electric Light Company in 1883.

The small steps taken by Edison and Swan led to the integrated technology of lighting that many of us in the Western world take for granted today. While it took many years to implement the necessary infrastructure for lightbulbs to be widespread throughout society, their introduction at the end of the 19th century created swift change. Society could now thrive at night, providing a safer environment for activities across the spectrum, including extending working hours and social activities into the night and facilitating navigation in the dark as a safer option outside of gas and oil lamps. Deaths were reduced in many industries and productivity increased, with numerous factories operating 24 hours a day as a direct consequence of access to lightbulbs – although this came at a cost to the factory workers.

Developments in the lighting industry did not stop once a practical design was found. Even though modern-day lightbulbs don't look so different from Edison and Swan's original designs, there was plenty of innovation. Tungsten filaments became the standard in the 20th century, for example, and today Light Emitting Diode bulbs (LEDs) are overtaking the traditional incandescent lightbulb in popularity – as predicted in the February 1963 issue of *Reader's Digest* by the developer of the first practical LED bulb, Dr Nick Holonyak Jr.

In the last decade LEDs have become one of the most common lightbulbs found in homes and offices, as they are more energy-efficient, more sustainable and cheaper, lasting thousands of hours longer than an incandescent bulb. Their design focuses energy on producing light instead of heat, meaning it doesn't waste anywhere near as much energy. Consequently, older and less efficient bulbs are being phased out with governmental reforms across the Western world, and modern lightbulb design has switched gears, prioritising environmental impact, electricity use and endurance.

Perhaps surprisingly, LED lightbulbs also took more than a century to become a viable lighting option within homes. The seed was sown as early as 1907, when the Englishman H.J. Round demonstrated that passing a current through silicon carbide would generate a small amount of light. This simple concept formed the basis of what would become the LED bulb. However, it took until 1962, when Holonyak Jr's LED was developed, for the LED bulb to become even remotely practical. His bulb produced just one colour – red – so it was viable only commercially for electronic components such as the time display in a radio. Nonetheless, it was an important step in LED lightbulb history, paving the way for more practical wavelengths of light to be produced, which better facilitated human activity. LEDs became a viable lighting option for everyday use at the end of the 20th century, with blue light made possible by the 1990s and white light by the new millennium.

But progress in the lightbulb industry hasn't always been so forward-looking. In 1925 a group of lightbulb companies known as the Phoebus Cartel came together to control the industry in Europe and the US, setting the standard lifespan of a lightbulb to 1,000 hours. This was much shorter than was actually possible at the time, with Thomas Edison's 1880 lightbulb capable of lasting 1,390 hours! This practice of engineering a shorter product lifespan or embedding failure into design to force consumers to repurchase is known as "planned obsolescence" and can be found in many industries in the modern capitalist culture. A bulb in California known as "the Centennial Light" demonstrates what could have been achieved had the Phoebus Cartel failed, receiving world records as the longest lasting bulb: it has barely been turned off since 1901, which is longer than the battle to achieve a practical lightbulb lasted, and *much* longer than any commercial lightbulb can last!

Both LED and incandescent lightbulbs illuminate the necessity of development, patience and collaboration in technological advancements. Often there is much more to a story than one individual inventor, and the history of the lightbulb showcases two centuries of development and the ideas of so many individuals.

Tesco's long-overdue range of plasters catered to darker skin tones.

PLASTERS

Sticking plasters, Band-Aids, adhesive bandages – whatever you call them, you almost certainly reach for this first aid kit staple when new shoes rub or a child falls and grazes their knee.

Humans have been dressing wounds to stem bleeding and fend off infection for millennia. Indeed, one of the earliest medical manuscripts – a 4,200-year-old Sumerian clay tablet – outlines the basic principles of cleaning an injury, making topical "plasters" from oil, mud and herbs, and wrapping with bandages. Adhesive dressings, too, have a long history. The ancient Egyptians employed honey not only for its sticky properties but for its natural anti-bacterial action. Yet the plasters we rely on today are a relatively recent innovation, dating back only to the early 20th century.

It was Earle Dickson, a cotton buyer for Johnson & Johnson, who came up with the idea of attaching a flat strip of gauze to the centre of a long piece of surgical tape, covering the entire thing with crinoline (a fabric then used for petticoats) to prevent it from sticking to itself. Earle's wife Josephine was prone to cutting her fingers in the kitchen; their invention enabled her to independently dress her injuries by snipping off a short section and applying it as needed.

"Inventor Earle Dickson's wife Josephine was prone to cutting her fingers in the kitchen; their invention of the plaster enabled her to independently dress her injuries."

Recognising the potential for a new product, Earle pitched it to his bosses. Johnson & Johnson were well-placed to bring plasters to market, having been mass-producing sterile surgical supplies for hospitals for the best part of three decades. Launched in 1921, the Band-Aid didn't exactly fly off the shelves, however; just $3,000 worth were sold that first year.

The first iteration was an 18-in. (45.72-cm) roll of adhesive bandage that still needed to be cut to size. Being handmade, it was relatively expensive, and perplexed would-be customers had to be shown how to use it. Johnson & Johnson hired travelling salesmen as demonstrators and launched an advertising campaign targeted at women, emphasising the ease and convenience of the Band-Aid.

The prosperous 1920s saw a boom in magazine and newspaper readership, giving retailers the opportunity to

A 1956 advert for Elastoplast. Adverts commonly targeted housewives, emphasising the plaster's convenience for treating minor injuries.

promote their brand-name products. Manufacturers of drugs and toiletries were particularly keen to get in on the action, spending more on newspaper advertisements than any other industry.

There were other factors, too, that contributed to the growing popularity of plasters. In the wake of World War I, householders were gradually becoming more aware of the importance of antiseptic wound care. Before antibiotics became widely available in the 1940s, even minor injuries could pose a risk to life from infection. It was during this era that first aid manuals came into their own. At a time when many relied on unsanitary strips of fabric to bandage cuts and burns, plasters fulfilled an unmet need for a sterile covering that could withstand the rigours of everyday life.

By 1924, Johnson & Johnson had introduced the familiar 3-in (7.62-cm), machine-cut adhesive bandage. Free samples were distributed in first aid kits specially created for the Boy Scouts of America. Intended to help the Scouts learn basic care skills and earn merit badges, the move familiarised families across the nation with the Band-Aid brand.

European manufacturers were hot on the company's heels, with German company Beiersdorf launching Hansaplast dressings in 1922 (later licensed as Elastoplast in the UK and Commonwealth). Tins of adhesive bandage were supplied to soldiers heading into World War II, further boosting their reach. By the 1950s, plasters had become a ubiquitous household item.

Makers continued to innovate as the baby boom generation grew up, creating plasters in various shapes and sizes, including square and circular patches. When, in 1956, Johnson & Johnson launched Stars 'n Strips – a range of bandages with bright, fun designs to appeal to children – the product was an overnight success. Over the years, new fabrics with features such as waterproofing, hypoallergenic and antibacterial properties were introduced.

Yet for decades after their invention (printed examples aside) plasters were available only in pink or light tan hues,

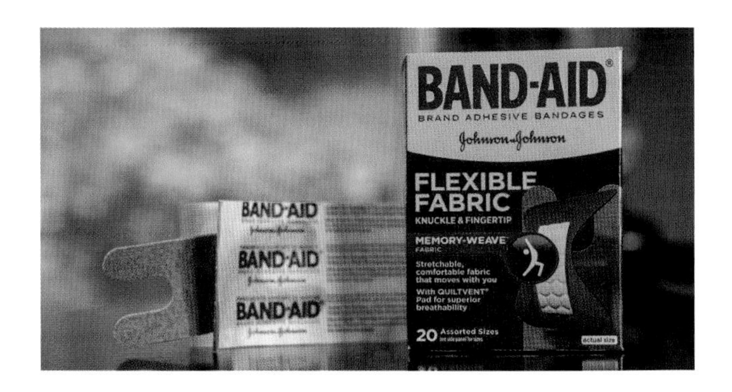

Plasters are now commonplace in every household first aid kit.

referred to as "flesh tinted" in adverts. Band-Aids and their European equivalents are just one example of many medical, fashion and beauty products designed primarily with white consumers in mind, meaning people with darker skin are often forced to use conspicuously pale bandages.

When some of the larger cosmetics brands began offering makeup for diverse skin tones in the 1980s, Johnson & Johnson developed Band-Aid Clear; a product that could be marketed to all for its "invisible protection". Yet the reverse side of the pad is – you guessed it – light beige.

Some 35 years later, in February 2020, Tesco introduced its own-brand fabric plasters in dark, medium and light brown shades. Promoted with the tagline "About bloody time", they were the first of their kind to be sold in a UK supermarket.

Until this point, plasters developed specifically for people of colour were available only via specialist retailers, often at an inflated cost due to smaller production lines. US brands including Soul Aid (1970s), Ebon-Aide (1998) and Tru-Colour (2014) paved the way for wider change. Within weeks of Tesco's announcement, Boots and Superdrug – the UK's largest pharmacy chains – launched their own ranges of plasters in a variety of skin tones.

Tesco claimed to have been inspired by a tweet from Dominique Apollon, of US racial equality advocacy group Race Forward, describing his emotional response to using a Band-Aid that matched his skin for the first time. The supermarket faced criticism, however, for allegedly copying the idea from two women of colour – particularly as its bold campaign failed to acknowledge the independent businesses that had pioneered diverse plasters.

Joanne Baban Morales, founder of Nünude, the British fashion brand that successfully petitioned to change the Oxford English Dictionary definition of the word *nude*, and Vivan Murad, creator of Swedish plaster company Skin Bandages, first pitched their collaborative product Skin Bandages By Nuditone to Tesco in 2018. The retailer showed no interest, but bought samples from Nuditone's Etsy store prior to developing their own range.

A century after introducing the world to plasters, Johnson & Johnson followed suit with OurTone Band-Aids in 2021. Having abandoned a similar line brought out in 2005, the company accelerated OurTone's launch in response to mass protests for racial justice following the May 2020 murder of George Floyd. While some welcomed the product as a small but significant step in the right direction, others saw it – noting the irony – as a "band-aid solution": a superficial fix to a much deeper problem.

ROVER SAFETY BICYCLE

Whether you commute by pedal power, enjoy bike rides at the weekend or remember practising with stabilisers as a child, you're likely to have experienced the freedom and thrill of a bicycle at some point. Although engineers have honed the design to make bikes lighter, faster, safer and easier to ride, the modern bicycle is fundamentally the same as the Rover safety bicycle developed by John Kemp Starley in 1885. Incorporating a geared-up chain drive to the rear wheel meant that both wheels could be the same size, eliminating much of the awkwardness, danger and discomfort of riding what is properly known as an "ordinary" bicycle, but is more commonly known by its nickname "penny-farthing".

The bicycle has been around since the 1700s, but the popularity of this specific kind of two-wheeled transportation didn't take off in the UK until the 1860s. Even then, bicycles were difficult to ride and likely to throw you off if you hit one of the many potholes or stones populating the poorly kept roads of the day. And cycling was very much a pastime for young, athletic, middle-class men. The Starley family experimented with many cycle designs and developments to make them easier to ride and more accessible to different genders, ages and levels of physical ability. James Starley's run-in with a patch of nettles inspired the invention of the chain-drive differential gear, allowing wheels to rotate at different speeds when cornering. Stable, safer tricycles, which could be ridden wearing the floor-length skirts of the period, were considered to be more suitable for women and Queen Victoria's order of two Starley Royal Salvo tricycles cemented their respectability.

"The Rover safety bicycle finally made cycling an option for all ages and genders."

However, it was the Rover safety bicycle that finally made cycling an option for all ages and genders. Its steel tube diamond frame made it lighter, and using a chain drive to power the rear wheel meant the rider no longer had the tricky job of both steering and pedalling via the front wheel. By running the chain around a larger gear on the pedal

With both wheels the same size, the Rover safety bicycle was a revolutionary design.

crankshaft and a smaller gear on the rear wheel hub, the safety bicycle could get rid of the excessively large, powered wheel of the ordinary or penny-farthing (sometimes as large as 1.5 m/5 ft in diameter), while maintaining an efficient use of pedal power.

The accessibility of the safety bicycle saw a surge in the popularity of cycling, although this was not immediate. By the mid-1890s, Britain was experiencing a bicycle boom to the point where demand outstripped supply. By opening up a means of relatively affordable private transport, cycling inevitably became very popular. For the majority of people who couldn't afford a horse and carriage, they were no longer constrained by railway timetables and routes. And compared to walking, cycling enabled people to travel longer distances at greater speeds.

Despite being invented around the same time, the motor car was an expensive and niche vehicle until Ford's mass production techniques made it somewhat more affordable from 1908. Additionally, cycling was much faster, with cars restricted by low speed limits, only rising to 20 mph (about 32 kmh) in 1903. The bicycle opened up the countryside to urban dwellers and enabled people living rurally to access towns.

However, it's important to note that not everyone could take advantage of the freedom offered by the bicycle. The best bicycles cost more than £30 (£3,600 in today's money), while second-hand bikes with outdated tyres would set you back somewhere between £2 and £5 (between about £240 and £600 today). Hire-purchase schemes enabled people of more moderate means to own bicycles, but realistically those taking up cycling were upper-working class or above.

The invention of the safety bicycle had a large impact on one group in particular: women. Like men, their travel had previously been restricted to foot, horseback or carriages, but with the added constraint that this was almost always accompanied. Transport by horse or carriage generally required a man or men to prepare the horse and carriage,

A group of men pose with their bicycles, c. 1900.

whereas the relatively light and self-contained bicycle gave women independence to travel without this reliance. The usually single-seater bike, and the seeming reluctance of chaperones to embrace cycling, also meant unmarried women were more likely to be able to travel unaccompanied. As well as creating more opportunities to spend time with the opposite sex unobserved, this offered a freedom rarely experienced by middle-class women, whose movements were otherwise closely controlled. Cycling also offered health benefits, by providing a form of active recreation.

Yet not everyone approved of the newfound freedom that cycling afforded women. Eliza Lynn Linton, a writer and ardent opponent of female cyclists, warned that this freedom put women in danger of "the intoxication which comes with unfettered liberty". In the US, where cycling had similarly exploded in popularity, many questions of decorum were raised, ranging from whether a female cyclist should always have right of way at a junction to whether a man should offer to help a woman with a flat tyre.

Equally controversial was what women should wear. The floor-length skirts of the period could easily get caught in the pedals, chain or wheels. So-called "rational dress" offered an alternative. Campaigning for clothing that was practical and comfortable, and did not harm women's health, the rational dress movement proposed knickerbockers (voluminous trousers gathered in below the knee), bloomers (baggy trousers gathered at the ankle, worn underneath a knee length skirt), and divided skirts as appropriate for cycling. It was popular in continental Europe and the US, but British women in rational dress were often viewed as immodest and unfeminine. Faced with verbal and physical attacks, rational dress was only really embraced by activists for broader female emancipation.

As a symbol and means of "unfettered liberty", the bicycle itself was also adopted by the women's suffrage movement. Able to cycle further than they could walk, campaigners used bikes to deliver leaflets. From 1907, activists like the Cycling Scouts of the London branch of the Women's Social and Political Union cycled out to the suburbs, other towns and rural villages to share their message of Votes for Women. Suffragettes additionally used bicycles in their militant activities as getaway vehicles to escape after planting bombs and setting fires. In 1914, Hilda Burkitt and Florence Tunks committed a series of arson attacks while on a cycling tour of Suffolk.

Around the world, the bicycle was seen as an emblem of women's emancipation. In 1896, American women's rights activist Susan B. Anthony stated:

"I think [the bicycle] has done more to emancipate women than any one thing in the world. I rejoice every time I see a woman ride by on a bike. It gives her a feeling of self-reliance and independence the moment she takes her seat; and away she goes, the picture of untrammelled womanhood."

So next time you jump on your bike, take a moment to think about how this two-wheeled machine, so little changed from the original Rover safety bicycle of 1885, has been a revolutionary force. And perhaps spare a thought for the ways it could transform society now and in the future.

Goblin D25 Teasmade, with detachable pottery teapot, 1955.

TEASMADE

Wouldn't it be great if your alarm clock could also make you a cup of tea? Well, for almost 2 million British homes, this dream was once a reality, thanks to automatic tea making machines like the teasmade. While their heyday came in the mid-20th century, these quirky contraptions were actually invented in the 1800s.

The first known automatic tea makers were created independently by two English inventors. There was Charles Walker, a gas engineer and inventor who was motivated by the merits of rising early and wanted to assist those who had "failed to acquire the habit", and Samuel Rowbottom, who also worked in electric and gas engineering. These early automatic tea makers were produced within months of each other at the end of 1891, and worked on a similar concept to the 1950s version pictured opposite. At a specified time, a trigger activated the tea maker's heater, boiling a kettle of water that had been filled in advance, until the steam forced the water out of the spout and into the teapot or cup. Once this happened, the alarm was activated, waking the user up to a fresh, hot cup of tea. There was one major difference between the 20th-century versions and the two earliest tea makers, however. Whereas the more modern versions used electrical heating elements, Walker and Rowbottom's designs heated water using liquid spirit and gas fires respectively, which was not particularly safe or quiet.

"At a specified time, a trigger activated the tea maker's heater, boiling a kettle of water that had been filled in advance, until the steam forced the water out of the spout and into the teapot or cup."

These early designs inspired similar but slightly altered tea-making alarms to spring up at the beginning of the 20th century, with excellent names such as "the Clock that Makes the Tea!", "Ron Grumble Early Morning Waiter" and the "Magic Kettle". In 1937, Britain finally had its first commercial automatic tea maker, or "teasmade", and the obsession with tea-making and waking devices was ignited. Available to purchase for your own home and thankfully powered by electricity, not a noisy and unsafe bedside fire, the teasmade shown on page 115 came from the company Goblin, as did the name we use today. (While originally a trademark of the Goblin company, the term *teasmade* is now considered a genericised trademark and used to refer to any automatic tea-making alarm clock.) Many different teasmades filled catalogues in the coming years, but Goblin became arguably the most well-known of the teasmade manufacturers, going on to produce various iterations throughout the early and mid-20th century, including the 1955 iteration pictured on page 112 – an example of the iconic teasmade style and design.

By the 1960s and '70s, the teasmade had reached its peak and was a common item in many UK households, as well as some Commonwealth countries. By the late 1960s over 300,000 were being sold every year, and by the 1970s they lived in around 2 million homes. Not just a novelty, the teasmade's popularity seemed to be the result of a combination of changing social conditions. By the early 1900s, the rise of mass production, coupled with improvements in travel and transport, meant that appliances like this could be made in large numbers and available in electrical catalogues nationwide.

A model shows off her Goblin teasmade.

The first Goblin teasmade in 1937 cost just over £5, roughly equivalent to £300 today. With average yearly UK earnings sitting around £150 in 1930, this certainly wasn't yet affordable to the masses, but would have been to affluent households who valued their morning cup of tea. As the 1930s came to a close, two-thirds of British homes had electricity, opening up new possibilities for plug-in devices. These factors also coincided with a change in the way household labour was performed. At the turn of the 20th century, wealthy households often relied on the domestic labour of live-in maids to perform tasks such as making breakfast. This practice started to decline, however, especially after World Wars I and II, when many women previously employed as maids began to find work in retail or clerical sectors, and those who continued to work in domestic service were less likely to live in their employer's home.

This only heightened as the country headed into the 1960s and '70s, with social and cultural changes allowing more women to join the workforce, making them less available to provide the unpaid domestic labour of the wives and daughters before them. Together, these factors paved the way for household appliances such as the motorised vacuum cleaner, the electric toaster and, indeed, the teasmade to enter the home and perform domestic tasks in the place of people. While a home well-kept by staff was once a status symbol, filling your home with the latest household gadgets became a new way to show off to your neighbours. With multiple models available from high street shops for around £15–£20 in the mid-1970s (roughly equivalent to £130–£220 today), the teasmade, for many Brits, was a worthwhile investment for the luxury of being woken every morning with a freshly brewed cup of tea.

The teasmade's reign over the bedside table was short-lived, however. By the 1980s and '90s they were largely seen as an out-of-date relic of a bygone era. Household trends will always come and go, often for unexplainable reasons, but one contributing factor here could be that many of these automated devices don't actually save you time. Most have additional cleaning or maintenance associated with them, and in the case of the teasmade, the time to make a cup of tea actually remains the same – you're just moving the task of filling the kettle to the night before instead of the morning.

Another flaw of the teasmade system is that if you want milk in your tea, it still needs to be collected from the fridge, or left out overnight. In a house that probably has more efficient heating today, unrefrigerated milk doesn't seem so appealing. Ultimately, the task of boiling a kettle for your tea is one that doesn't require excessive labour, and perhaps people realised that it wasn't worth buying a whole new device to automate it. In today's world, an automatic tea-making alarm clock doesn't seem such a perfect fit as it might have done in mid-20th-century Britain. This is still a nation of tea drinkers, but the arrival of international barista coffee chains or artisan coffee shops have many Brits craving elaborate coffees instead, and globally many people choose to invest in machines that can recreate these more complicated hot drinks at home. The alarm clock itself, even without an attached tea maker, is also somewhat obsolete, with most people relying on their mobile phone alarms to wake them up.

The original "Teasmade" automatic tea-making machine released in 1937.

The teasmade's hold over British bedside tables may be long gone, but the spirit of inventing household appliances to help with domestic labour isn't going anywhere. With busy lives, many people choose to delegate daily tasks to automatic or remotely activated appliances. In certain homes today, you might find vacuum cleaners that move by themselves, cupboard buttons that can automatically purchase replacement toiletries (now discontinued), or a whole host of electronic appliances that can now be activated remotely from a smartphone – even sometimes the kettle.

"Filling your home with the latest household gadgets became a new way to show off to your neighbours."

TV SETS

In 1971, 10 per cent of UK homes had no indoor toilet, no indoor bath, nor shower. That same year, 91 per cent of UK homes had a TV. Such uptake is startling, especially as the first commercial television only became available in 1928. But why had television been prioritised over essential household sanitation?

Researchers have found that TV's popularity goes beyond its ability to entertain. TV can create escapism, provide psychological benefits and emotional fulfilment. One study found watching TV slowed our reaction time to secondary tasks, allowing our brain to be captivated and distracted from our everyday life. In this study the secondary task was simply pressing a button, but in real life secondary tasks could be knitting, chatting, cooking – anything done at the same time as watching TV. The time taken to do these tasks increases when TV is there to distract. This phenomenon is explained by a theory called the "Limited Capacity Model of Motivated Mediated Message Processing" or LC4MP. As LC4MP's name suggests, humans have a maximum amount of information they can cognitively process at a time. And watching TV, which requires lots of different types of cognitive functions, absorbs a significant amount of energy, preventing us from being able to focus on the other things we're doing. Through this act of distraction, TV also allows us, if just temporarily, to forget about our worries.

Other researchers have found that watching TV is restful. It was ranked 10th in the "Rest Test", a large-scale and international survey on restful activities undertaken in 2016. Humans seek activities that require little effort while simultaneously enabling us to recharge our batteries. And in this case, watching TV provides passive rest – your brain is distracted while your body relaxes.

"Watching TV, which requires lots of different types of cognitive functions, absorbs a significant amount of energy, taking our minds off things."

Watching TV has been shown to have an emotionally fulfilling dynamic. Repeatedly watching the same content can create emotional stability and a sense of control over one's life – the ability to turn the TV on at a certain time, to tune into an expected programme, or rewatch a favourite series (again and again). Rewatching our beloved content, whether that be *Friends*, *Game of Thrones* or *The Sopranos*, can also create the sensation of experiencing something a

John Logie Baird's original experimental television apparatus.

little new. This is because of our selective memory – we suddenly notice little parts of a storyline that we missed or overlooked before. It is this subtlety that ensures we're still entertained; something new is embedded in the familiar, balanced by the reassuring knowledge that the ending won't change.

Studies have also suggested that a parasocial dynamic (a one-sided relationship) is created with certain TV content. Viewers can become invested in and affected by the fictional or real people they see on screen. This relationship is built week after week, or episode after episode, leading to a connection and emotional fulfilment similar to that provided by real life friendships, family and romances. Such emotional attachment is proven by countless online lists, such as "The 17 major TV character deaths we're still mourning". We feel shocked, a real loss, over the deaths of certain TV characters, or true joy over their (fictitious) successes.

Yet the physical object, the technology and the complex emotional and psychological effects offered by TV are only 100 years old. During the 1920s there was fierce competition in the race to invent television, with makers working all over the globe. There was Japan's Kenjiro Takayanagi, who used a system of scanning and electronic signals to project an image in 1926, America's Philo Farnsworth, who is thought to have created the first electronic television transmission in 1927, and Britain's Isaac Shoenberg, who created the Marconi-EMI system in 1936. But it is John Logie Baird, a Scottish inventor, whose "televisor" ensured him the title "grandfather of television". But the televisor is distant from the sleek objects we recognise as TVs today and the impactful content we're used to seeing on them. It was exceedingly homespun, was made from a hat box, tea set, knitting needles and bike light lenses. Baird converted these pieces into the televisor: a large disc with 30 holes filled with lenses in a spiral, a neon lamp receiver, and a picture scanning transmitter. As the disc turned, the lenses scanned what was in front of it. These scans happened in vertical lines and were turned into electrical pulses by shining them onto a photocell – a resistor that changes electric pulse depending how much light is on it. The pulses were intensified and transmitted, causing the neon light to shine at different intensities (brighter and darker). This light was sent through a second disc with spiral holes. These holes reversed the scanning process, projecting the neon light when it was brighter and darker, creating a black and white silhouette of the subject being filmed, or "televised". The first object to be televised was a puppet's head. Initially the image appeared only faintly, but as Baird perfected his experiments, the puppet became clearer and clearer.

Even before Baird's first fully televised image, he demonstrated an earlier iteration to the public on 25 March 1925, in Selfridges Department Store, London. While this iteration was even more shadow-like, the choice of Selfridges perhaps established the commercialisation of TV before the technology was actually realised. In his 1985 book *Amusing Ourselves to Death*, American writer Neil Postman noted that there is nothing, no lack of time nor money, that has ever gotten in our way of watching TV. And this is true from the beginning. Baird's televisor Model B, which went on to be the first mass-produced, commercial unit, sold for £40 in 1928 (today £3,215) – a huge sum then and now.

Even when TV became mainstream in the 1950s, in every part of the world, it remained costly. In 1951 a standard black and white TV set could cost £80 (well over £3,000 now). In 1970s Britain, an average black and white TV set cost £69 (almost £1,000 today). People today can spend tens of thousands of pounds on a TV, but most people in the UK spend somewhere between £300 to £500 – which is still not cheap.

This brings us back to the fulfilling dynamic of TV and why money really is no object. Content has always been just as important as the physical technology. The BBC's role in UK TV was crucial from the start. First, by providing Baird access to their London transmitter, which allowed Baird to run a programme of broadcasts. Then the BBC went on to produce the first play ever broadcast on a Baird TV – Luigi Pirandello's *The Man with the Flower in his Mouth*. The BBC mantra to "entertain" and "inform" became guiding principles. They did so with *Monty Python* (on BBC One from 1969) and massive sporting events like the "Fight of the Century" of Muhammad Ali vs. Joe Frazier II (watched by 21.12 million people on BBC One). Since then, TV content has expanded, and content providers have become more numerous and varied. Take Netflix's niche content categorisation or "micro-genres" – "Wine and Beverage Appreciation" or "Cerebral Scandinavian Movies", for example. Netflix is also known to experiment, bringing together elements of random TV genres to create new and unexpected shows, such as *Stranger Things*, which leaned heavily on a wacky mix of science fiction, childlike fun from its young cast and nostalgia for the '80s. Regardless, TV content has been responsible for some of the most powerful moments in the history of popular culture, from the broadcasting of the moon landing to the Ross and Rachel kiss.

The continued success of television is in part due to the way TV watching habits have evolved through time. We still love it, but not necessarily on the box. One in eight people in the UK watch TV on their phones at least once a day, and the proportion of those watching traditional TV (in front of the set in the lounge) dropped from 83 per cent in 2021 to 79 per cent in 2022. This shift has been enabled through the development of physical technologies, from phones to tablets, as well as the move away from traditional live television broadcasting to streaming services. Regardless, the TV as a physical object has endured. In 2022, 96 per cent of UK households had a TV. It would be almost strange to go into a British person's lounge and not see a TV set, with the sofas and chairs arranged in preparation for viewing. The TV set is still popular for watching festivals and big events, often with other people. The NFL estimated 21.8 million people watched the Super Bowl live broadcast in 2023, often considered the biggest annual sporting event in the US.

From the silhouette of a puppet to zeitgeist moments, TV has come a long way in terms of technological and content development. It has come even further in terms of infiltration into our lives – the sheer amount of time we spend watching it. We like to be entertained, but it is perhaps all the other deeper reasons, from emotional fulfilment to restfulness, which get us hooked. In this sense, perhaps, it is time and money well spent.

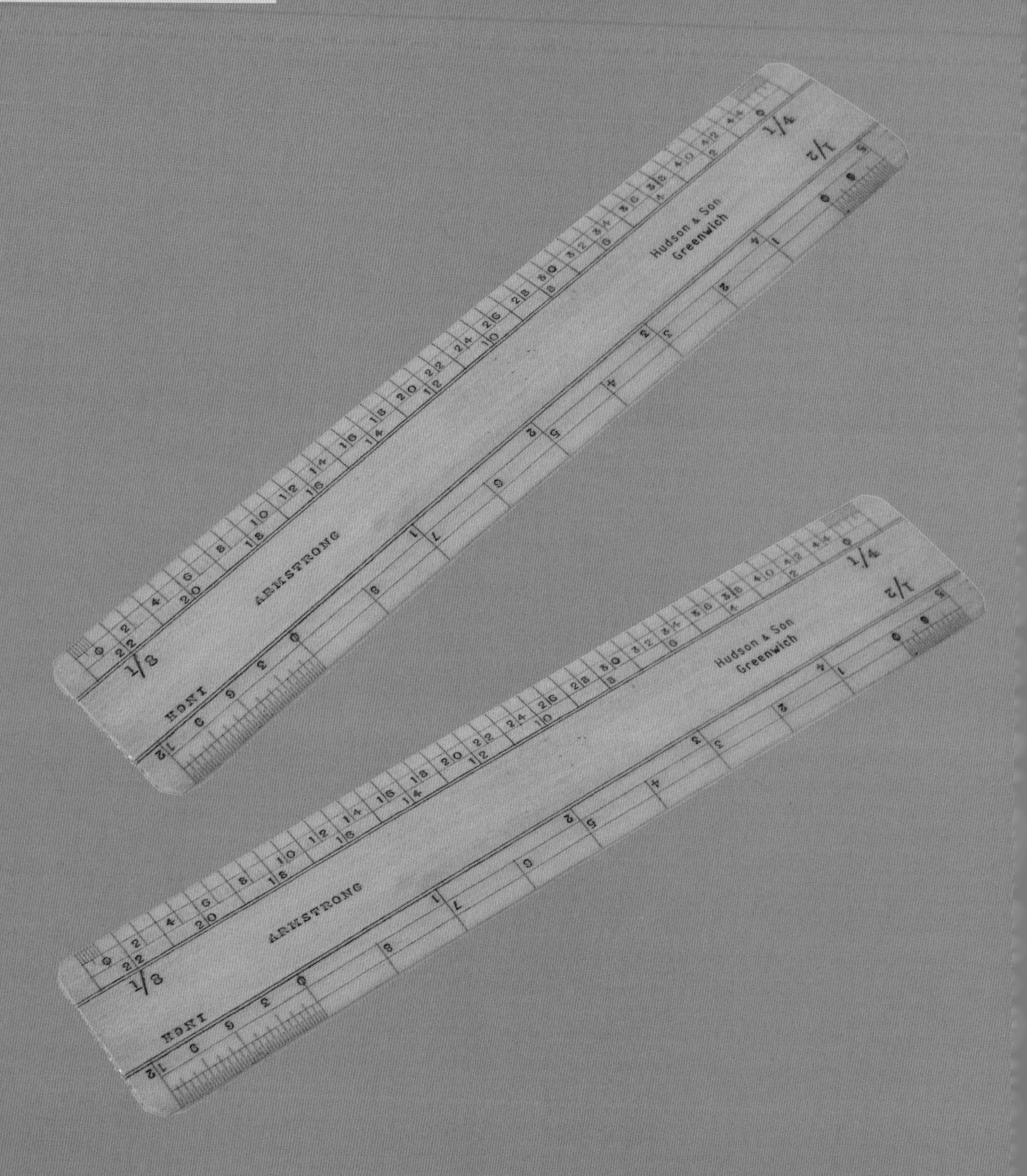

Wooden ruler marked "Armstrong", made by Hudson & Son, Greenwich.

RULERS

The ruler has always been a pencil case essential. There's often one lying around somewhere, forgotten about in an old drawer, pot or tool kit, until the odd moment when it comes in handy. This rather mundane piece of stationery has played two key roles in history.

Rulers, in their most rudimentary form, are any rigid item used to measure length. The earliest tools used for measuring length were body parts like arms, palms, fingers and feet. However, given that the length of these varied from person to person, they were of limited value for communicating measurements. As the civilisations and writing systems of ancient Egypt and Sumer were developing around 5,000 years ago, there was an increasing need for standardised measures so that instructions could be passed between people and understood. The first objects we would recognise as rulers were rods marked with standard lengths. The oldest surviving example of these is a cubit rod dating to 2650 BCE from the ancient city of Nippur in modern-day Iraq. The cubit was a measurement based on the distance from the elbow to the tip of the middle finger, and it was subdivided into the smaller units of palms and fingers. Its precise length varied from place to place, but the use of standardised measures was key to trade and construction.

"There was an increasing need for standardised measures so that instructions could be passed between people and understood."

The history of metrology, the scientific study of measurement, has been at the root of advances in technology, science and society over thousands of years. Indeed, measurement is a fundamental component of the scientific method: measuring the output of experiments enables us to test and prove hypotheses. The ability to quantify anything and everything is key both to our understanding of the world around us and to our ability to manipulate it. And so, whether it's a kelvin (named after British mathematician Lord Kelvin), a joule (named after English physicist James Prescott Joule), a watt (named after Scottish inventor James Watt) or a newton (named after English scientist Isaac

Newton), it is no surprise that one of the peaks of scientific achievement is to give one's name to a unit of measure. (It is also an indictment of the history of science that all the specially named units in the International System of Units are named after white men.)

It is testament to the importance of measurement in our lives that they have played their part in revolutions throughout history. Abuse of weights and measures by rulers (the other kind), especially where taxation and trade were involved, eroded trust in authority. Democratic movements also recognised the importance of standardised weights and measures for facilitating trade. Article 1, Section 8 of the US Constitution gave congress the power to "fix the standard of weights and measures" for exactly this reason. Clause 35 of the Magna Carta, the English charter of 1215, specified the need for a standardised measure for wine, ale, corn, cloth and weights across the kingdom. Meanwhile, the French Revolution was fundamental in starting the journey towards our modern system of measurement.

On the eve of the French Revolution in 1789, France had hundreds of different units of measure, whose definitions varied between each town and even between different professions. It was a nightmare for trade, sowing distrust between traders. In 1790, in the wake of the revolution, the Académie des Sciences appointed a panel to investigate the problem. They recommended a system of measurement based on physical constants that could always be referred to, proving that the unit had not changed. For the metre, they chose the distance from the North Pole to the Equator, measured at the Paris meridian, divided by 10 million. Their unit of weight – the kilogram (kg) – was based on the weight of a volume of water, with dimensions based on the unit of distance that they had created. This gave rise to both the metre (m) and the gram (g), the metric units of length and mass that dominate the world today. This evolved over the next two centuries into the commonly used International System of Units (SI), in which every unit is derived from physical constants. Indeed, the system continues to evolve; the most recent update in 2019 sought to define the kilogram in terms of the Planck Constant.

"Democratic movements also recognised the importance of standardised weights and measures for facilitating trade."

The ruler's place in this history begins in the earliest days of measurement. But as measuring devices go, the ruler is even more special because it has a dual purpose – it can also be used to draw straight lines. The ability to draw straight lines and measure them at the same time may not sound particularly groundbreaking today, but it is fundamental to a technology that has immeasurably shaped our lives – technical drawing.

Technical drawing is, in its essence, a language. Often employed in architecture or engineering, it is the act and discipline of composing precise and detailed drawings that visually communicate how something functions or is constructed. Once information about the three-dimensional world is converted into a two-dimensional format, it can be transported and stored just like a book. But just as a picture can paint a thousand words, a technical drawing stores information far more efficiently than any linguistic description ever could. Moreover, unlike photographs, technical drawings can capture that which does not yet exist, enabling us to translate ideas into physical form. Draughting is an exercise in science, maths, invention and art, all on the path to creating something new.

The world's oldest surviving technical drawings are almost as old as the ruler. They appear on statues of Prince Gudea of Lagash from around 2150 BCE. Only 85 miles (135 km) and 500 years separate the statues from the Nippur cubit rod. The drawings depict plans of Prince Gudea's palace, including a scale of measurement. No working drawings from this period exist today, because the medium on which they were produced did not survive, so we rely on these statues as evidence that the activity of architectural drawing was already a well-established skill more than 4,000 years ago. The same is true from the era of Roman and Greek civilisation. However, books like *De Architectura* by the renowned Roman architect Marcus Vitruvius Pollio, which survived thanks to copying by monastic scribes, still provide significant insight into the draughtsmanship skills of the era's architects.

Technical drawing has a long history, but its significance grew exponentially during the Industrial Revolution. The volume of drawings being produced also exploded, all drawn with the aid of compasses, set squares and, of course, rulers. The value of technical drawing lay in the ability of engineers to communicate their ideas to their colleagues. Without technical drawings, engineers had to describe their designs in words and demonstrate them with models and prototypes – a laborious process that was prone to error. Technical drawings were a far more efficient means of communication, facilitating the specialisation

Technical drawing offers engineers the ability to communicate their ideas efficiently.

A train-shaped ruler created to promote Virgin Trains.

of roles; the task of closely monitoring the manufacturing process could be delegated to supervisors, managers and craftspeople, freeing up engineers to focus on their designs. Drawings could also be transported, ensuring that manufacturing could be distributed to distant locations or subcontracted to accelerate manufacture. When the process of blueprinting was introduced in the mid-19th century, engineers were also able to make quick and cheap copies of their drawings for the first time.

Technical drawings could also be stored. This made the repetition of designs far easier, moving manufacturing away from bespoke artisanal products towards the standardisation of products and their components. Technical drawings were stored by patent offices, allowing engineers and inventors to protect their designs from plagiarism, and incentivising innovation. Repairing, changing and updating objects is also far easier when their original drawings are available for reference. Thousands of engineering drawings at the National Railway Museum are accessed every year to assist in the repair and operation of heritage railways across the UK, and Network Rail, who maintain the UK's railways, retain more than 6 million drawings to aid in the maintenance of the railways.

Technical drawing was an incredibly valuable skill. It was a cornerstone of the Industrial Revolution, enabling rapid advances in technology, mass manufacturing and innovation. However, in the context of its long and illustrious history, the demise of technical drawing was rapid. The first computer-aided design (CAD) systems were developed in the 1960s, and by the end of the 20th century the advantages of CAD software were clear. CAD allows engineers to create from their designs far more rapidly, interacting with other software systems to operate automated machinery. In a few decades, technical drawing and the draughtsperson's kit of rulers, set squares, compasses and other tools virtually disappeared from industry and engineering. Even so, the ruler will likely live on in many drawers and tool kits for years to come.

THE CONTRACEPTIVE PILL

The contraceptive pill has been hailed as one of the most radical and transformational medical innovations of the 20th century. Its legendary status is such that it is the only medicine to be known by the simple moniker "the pill". Approved for use in 1960, it was the first hormonal birth control medicine. Made with synthetic versions of the sex hormones oestrogen and progesterone, the pill works by preventing ovulation and thickening mucus in the neck of the womb to block out sperm.

Credited as a catalyst that heralded the sexual revolution of the Swinging Sixties, the pill provided the power to determine one's own reproductive future. As this new birth control method was nearly 100 per cent effective, for the first time it was possible to have sex without fearing unplanned pregnancy. With the option to delay starting a family until later in life, more people were able to pursue careers and higher education.

The positive impact of the pill on lives across the world cannot be understated. But how did it come into being? The story of this little tablet is one of feminist innovation, exploitative and unethical human experimentation, and disruptive campaigning for more stringent drug safety.

The idea of a contraceptive pill was first conjured by Margaret Sanger, an American birth control advocate, early feminist and sex educator. Sanger believed that every woman should be "the absolute mistress of her own body". She argued that women could not be emancipated until they had a way of controlling their own fertility. In 1916, she founded the Planned Parenthood Federation of America, the first birth control clinic in the US.

"The idea of a contraceptive pill was first conjured by Margaret Sanger, an early feminist."

Sanger fought for birth control at a time when it was forbidden in America. The 1873 Comstock obscenity laws had prohibited the distribution of contraceptive devices and sex education literature. Similarly restrictive views

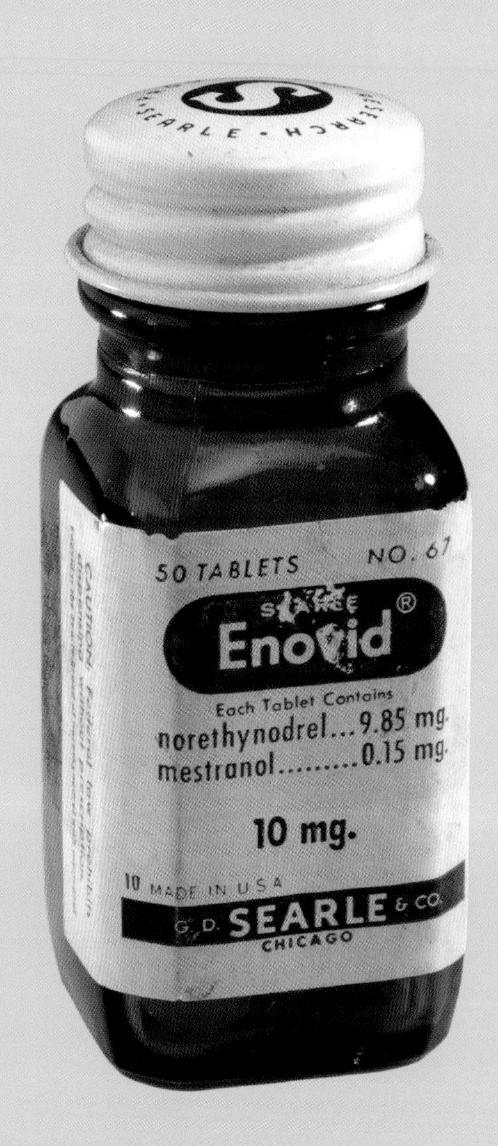

A bottle of Enovid pills, the first contraceptive pill to be approved by the US Food and Drug Administration.

governed approaches to family planning across many countries in Western Europe. Disregarding the law, Sanger sought out funders and scientists to make her dream of an oral contraceptive a reality.

The endeavour was financed by Sanger's longstanding friend: suffragist and heiress Katharine McCormick. With funding secured by the early 1950s, Sanger turned to Harvard University scientists to take the project forward – namely biologists Gregory Pincus and Min Chueh Chang, and obstetrician John Rock. All had been conducting research into fertility and had used synthetic sex hormones in animal studies. Although Chang, Pincus and Rock would successfully develop a contraceptive pill, the methods employed for testing were unethical and dangerous by modern standards.

Pincus began a small-scale human trial of the contraceptive pill at Worcester State Hospital, Massachusetts, in 1954. Under the pretence of investigating a possible tranquillising side effect, he recruited 16 female psychiatric patients into the study. These women were experiencing chronic psychosis, so were unable to give informed consent. While researchers obtained permission from the patients' families, relatives were deliberately misled about the trial's true purpose, which was to observe the long-term effects of hormonal birth control on ovulation. After administering the pill, laparotomies – a procedure where the abdomen is cut open – were performed to see how the drug had affected the patients' reproductive organs.

"In a pre-trial study, medical students under-went daily vaginal smears, monthly ovary biopsies and, occasionally, laparotomies."

The preliminary data was promising, but the pill needed to be trialled on a larger group of fertile, sexually active women before it could be approved by the Food and Drug Administration (FDA). In a letter to Sanger, McCormick described her desire for a "'cage' of ovulating females to experiment with". Knowing it would not be possible to conduct clinical trials in America due to the aforementioned anti-birth control laws, the team turned their attention to the US colony of Puerto Rico.

The American authorities that governed Puerto Rico believed overpopulation was the primary cause of its widespread poverty. The rise of eugenics – the scientifically and morally flawed belief that human populations can be improved by allowing only certain people to reproduce – led to the legalisation of barrier method contraceptive devices and a state-run sterilisation campaign in the 1930s. Not only was contraception already entrenched in Puerto Rican society, but its island status ensured that the population was contained and relatively stationary. Pincus and Rock had found their "cage".

In a pre-trial study, the pair recruited medical students from the University of Puerto Rico. While taking the pill, the students underwent daily vaginal smears, monthly ovary biopsies and, occasionally, laparotomies. Unsurprisingly, this degree of invasiveness proved too much for many, and half withdrew.

Circular pack of Norlestrin, an early contraceptive pill.

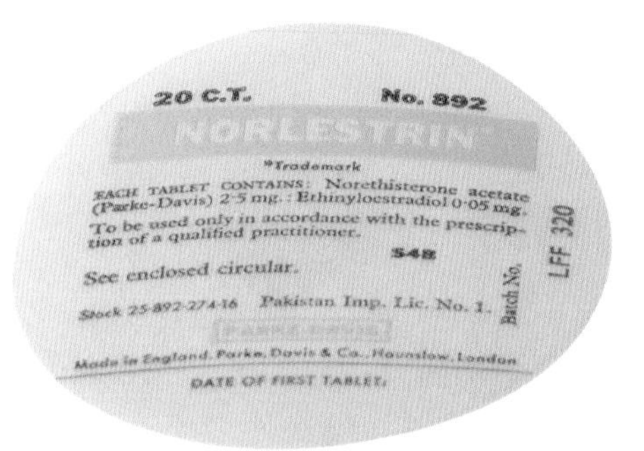

The official field trial of the pill began in 1956. Puerto Rican women from poor communities were given extremely high doses of hormones, without being told the treatment was experimental. Informed consent is now considered a critical requirement for ethical research, but clinical trials were relatively unregulated during this period. While the 1947 Nuremberg Code stated that explicit voluntary consent should be given by test subjects in human experimentation, it was not legally binding.

Many of the participants reported unpleasant side effects, ranging from blood clots to vomiting and depression. Pincus and Rock disregarded these reports, suggesting the women's symptoms were psychosomatic. Three died during the trial, but autopsies were not conducted, and the scientists claimed the deaths were unrelated.

The first contraceptive pill, Enovid, was approved by the FDA in 1960. It was a convenient, small tablet, taken daily for 21 days with a 7-day break. It became clinically available for married women in most Western countries in 1961. The roll-out for unmarried women occurred a few years down the line: 1967 in the UK, and 1972 in the US. By 1965, 6.5 million Americans were on the pill. In a twist of dark irony, its success was not mirrored in Puerto Rico; the retail price of $11 was too expensive for most to afford.

The first version of the pill delivered hormones in much higher dosages than those prescribed today. Enovid contained around 100 times more progestin (the synthetic version of progesterone) and more than two thirds the amount of oestrogen found in modern pills. As a result, first-generation pills caused severe side effects and posed rare but very serious cardiovascular risks to users.

In 1969, Barbara Seaman, an American women's health journalist, published

The Doctors' Case Against the Pill. Her critique of the drug's safety prompted the Senate Pill Hearings, an assembly of the US Senate and scientific experts to discuss the contraceptive pill. On 14 January 1970, women's health activists advocating for more transparent communication of the pill's side effects attended the first day of the hearing. Members of the D.C. Women's Liberation collective disrupted the hearing to protest the Senate's refusal to hear testimony about experiences of using the pill. Alice Wolfson and others shouted questions to the male panellists from their seats in the audience. They demanded to know: "Why are 10 million women being used as guinea pigs?"

In 1975, Wolfson and Seaman co-founded the National Women's Health Network (NWHN). The NWHN held a demonstration at the headquarters of the FDA on 16 December 1975, calling for patient packaging inserts (PPIs) to be included in the packaging of all oestrogen-containing drugs. The network argued that the disclosure of side effects was essential for women to give informed consent to treatment. The FDA listened to the protestors' demand, and since 1978 all packages of the pill have come with PPIs. This legacy is noteworthy, as it paved the way to PPIs for all medications.

In the years that followed, researchers were pushed to develop safer, lower dose formulations of the pill. Although conversations around the merits and drawbacks of hormonal contraception are ongoing, the pill remains the most popular method of reversible contraception in Europe and North America.

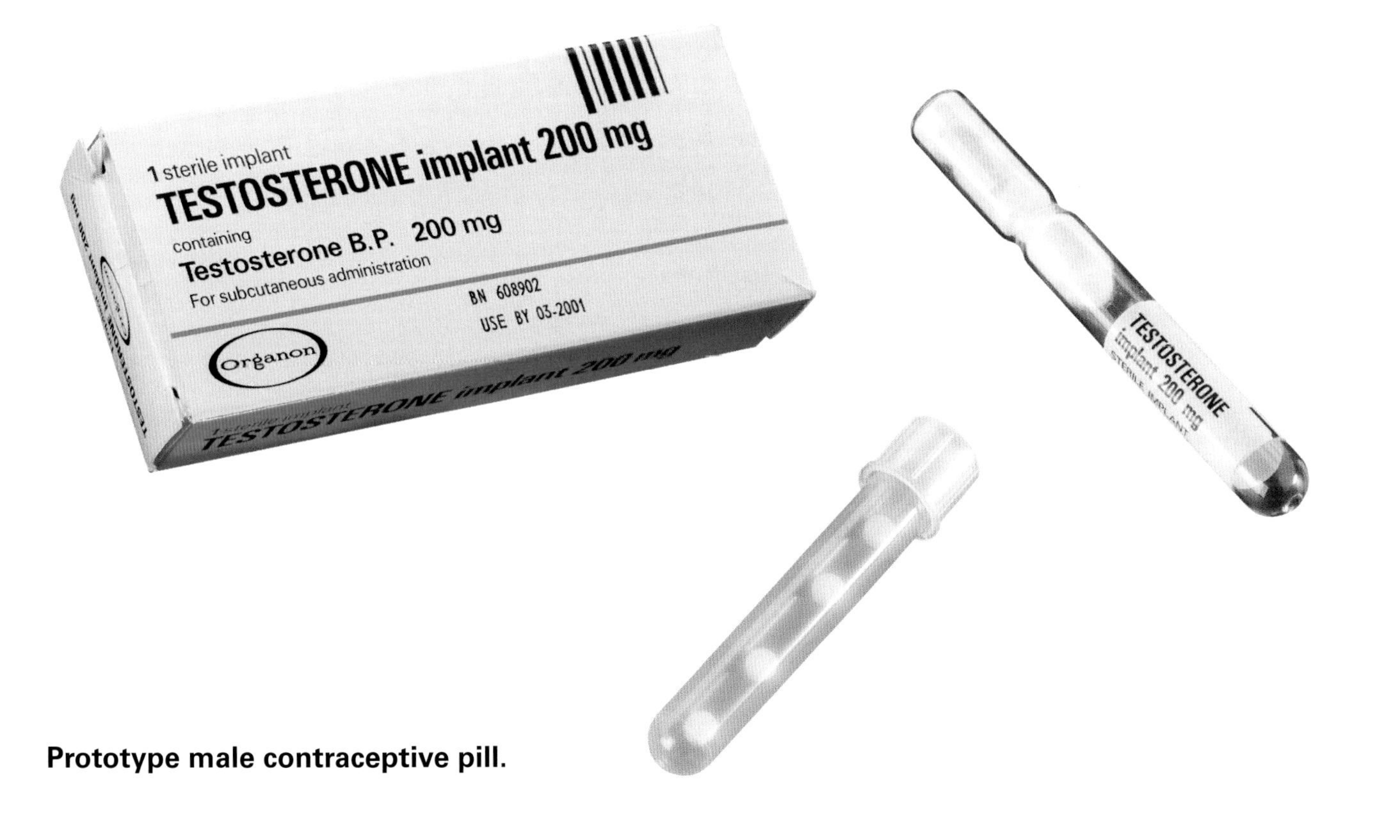

Prototype male contraceptive pill.

FRIDGE MAGNETS

Are you a memomagnetist, a thuramgist or a ferrovenirist? If your kitchen is covered in souvenirs and mementos, you may be all three, as each term has been proposed to describe people who collect fridge magnets.

A relatively recent creation that harnesses an ancient natural phenomenon, the fridge magnet's popularity has exploded since its inception in the 20th century. Love or hate these kitsch kitchen decorations, their affordability, convenience and light-hearted novelty mean they can be found for sale in gift shops and adorning kitchens worldwide.

The existence of the fridge magnet relies on two things: first, that the object itself is made of a relatively strong, permanent magnet; and second, that the fridge's outer shell is made of a magnetic metal. Whether a material is magnetic or not depends on its atomic structure. Metals such as iron, nickel, cobalt and steel are usually attracted to magnets, for example. While not every fridge throughout history has been made of magnetic material, most have, and today a fridge's outer layer contains sheets of steel. Some magnets are permanent magnets, which generate their own magnetic fields constantly, while some require an electrical current to generate a magnetic field. The souvenirs we adorn our fridges with are permanent magnets, meaning they are constantly producing the force required to stick and not fall to the floor.

Artificial refrigeration began in the 1750s, with early examples being used mostly in industries such as health, ice making, and food production and distribution. Household refrigerators were developed in the 1910s, but didn't become widespread in the homes of the general public until closer to the 1930s. The fridge was a revolution in home food preparation, keeping perishables safe and free from bacteria for far longer than the ice boxes that were previously used. At a similar time, scientists were working on a way to imbue materials with permanent magnetism. While these two innovations don't cross paths in this story until a few decades later, the invention of ferrite in 1930, a magnetic compound cheaper to produce than previous permanent magnets, paved the way for the affordable kitchen decorations we know and love today.

"The invention of ferrite in 1930 paved the way for the affordable kitchen decorations we know and love today."

The LHC "coolest magnets" fridge magnet.

"In the 1970s, William Zimmerman's company Magic Magnets was creating fun, novelty magnetised cartoons."

As a novelty object, the history of the fridge magnet is less well recorded, but it begins with magnets used more for function than for fun. In the mid-20th century, small magnets were often attached to letters and used in educational settings, allowing them to be stuck, moved around and then removed when you were finished. Continuing with their functional beginnings, in the 1960s American mould-maker Sam Hardcastle was approached by the space industry to create magnetic numbers and letters that could be used on large visual tracking charts. These new magnet shapes differed from the previous educational letters in that the whole object was magnetic and flexible, and not simply a letter attached to a separate magnet. This made them even better at sticking and staying.

Developing his moulding and painting techniques, Hardcastle was soon able to make magnets in a multitude of shapes and colours and saw the potential for these beyond the space industry. He initially started an advertising agency, making promotional magnets for companies, and later moved into souvenirs, designing speciality fridge magnets for each of the 50 US states. Hardcastle wasn't the only person creating small magnetic objects at this time, though. In 1954, a patent had been filed by inventor Oren Whitwell for a magnetic memo holder that could be stuck to any suitable metal surface, and small adverts for fruit-shaped magnets could be found across various American newspapers in 1962. In the 1970s, William Zimmerman's company Magic Magnets was creating fun, novelty magnetised cartoons. The history of the fridge magnet is unfortunately not extensively recorded, but it's clear that fridge magnets like the ones we are familiar with today were starting to spring up in different places across the mid-1900s. In the span of less than a century, however, the fridge magnet has gone from its early inceptions in a few niche settings to an almost ubiquitous object. They are sold at tourist sites and gift shops in the furthest corners of the globe, and are available in more shapes and designs than you could imagine.

Fridge magnets today are mostly used as decorations and souvenirs. Representing trips taken, companies worked for, family photos or motivational quotes, the fridge magnet allows people to display their loves, interests, and memories in plain sight. And for people living in rented or temporary accommodation, who aren't able to decorate the walls or hang photo frames, the impermanence of the fridge magnet offers a changeable decorative display, without the risk of a lost deposit.

The fridge magnet still has functional uses, often tasked with holding up wedding invitations, shopping lists or notes to return important calls. The fridge is the perfect location for these reminders, as most people visit it multiple times throughout the day. Fridge magnets could even have more serious uses than this. In a *Guardian* article from 2012 titled "Cold Comfort: in case of emergencies, please contact my fridge", the author speculates about whether the fridge is the ideal location to display important personal information and instructions to be found by emergency service workers, family or friends in times of need.

Functional or not, the fridge magnet frequently divides opinion. For some, the idea of filling your kitchen with brightly coloured, tongue-in-cheek plastic mementos from tourist shops is tacky and undesirable. For a few magnet fans, however, the joy of collecting these objects snowballed. At the start of the 1990s, fridge magnet collector Marlou Freeman was looking to downsize her home and sell all 2,300 magnets in her possession. The newspaper advert caught the eye of Manhattan gallery owner Alesh Loren, who claimed Marlou had "put together a masterpiece of American popular culture in miniature", and instead displayed her collection on 64 old fridge doors in an exhibition for members of the NYC art scene to marvel at.

The world record for fridge magnet collections goes to the late Louise Greenfarb, who owned an incredible 35,000 non-replicated magnets at the height of her collection. Neither Marlou nor Greenfarb had initially intended to create notable collections; instead, they fell into the habit, collecting so many magnets that they spilled off the fridge onto other kitchen appliances, and finally across the house, onto purposely installed metal surfaces.

The magnet pictured on page 131 is the perfect example of a fridge magnet's ability to turn a serious subject into something playful. The Large Hadron Collider (LHC) at CERN in Switzerland is the world's largest and most powerful particle accelerator. It speeds up and smashes together particles in an effort to understand what the universe is made of, how it was formed, and what might lie in its future. With its pun that the LHC dipoles (magnetised poles separated by a distance) are the "coolest magnets", this fridge magnet sold at the CERN gift shop makes reference to and plays with the function and structure of the LHC, itself a series of thousands of enormous magnets that are supercooled to -271.3°C (-359°F).

Magnetism has been harnessed for a multitude of purposes, and magnets can be found all around us in ways we might not have realised, from MRI machines and medical treatments to clothing and jewellery; from factories and machinery to computers, trains, toys and many household appliances. Of all these uses though, perhaps the most joyful and most frequently seen example is the humble fridge magnet.

Fridge magnets from around the world.

TOOTHBRUSHES

If you think hunter-gatherers must have had worse breath and teeth than us, you may want to think again. It's hard to say if our ancestors had fewer cavities than we do today, but one thing is for sure – a carbohydrate- and sugar-filled diet causes more tooth decay. Around 10,000 BCE, during the First Agricultural Revolution, the farming of grains and sugar led to a diet that required more regular teeth-cleaning to help protect against the harmful acids in sugar. The toothbrush, in collaboration with toothpaste, has become an integral part of our daily hygiene routine. But it's more than just having fresh breath and white teeth; after a week without brushing, our enamel starts to break down.

The story of the toothbrush starts with the *miswak*, a teeth-cleaning twig made from the *Salvadora persica* tree, native to Africa, West Asia and India. The miswak is made from the roots, stem and twigs of the tree, which are collected and cut into holdable sticks. One end of it is shaved to reveal the fine fibres that work between the teeth when chewed on. The earliest evidence found for its use is in 3500 BCE Babylonia, and 3000 BCE in an ancient Egyptian tomb. To this day, it has a large religious significance in Islam as a Hadith suggests making regular use of miswak, and it is widely used across West Asia, South Asia and North Africa.

"The story of the toothbrush starts with the *miswak*, a teeth-cleaning twig made from the *Salvadora persica* tree."

In the Tang dynasty of ancient China, between 618 and 907 CE, the first constructed toothbrush made of a bamboo or bone handle with boar's hair for bristles emerged. This toothbrush reached Europe only in the 17th century, when European travellers brought it back from China. Toothbrushes remained a niche product in Europe until the late 1780s when English entrepreneur William Addis created Addis, a company that would be the first to mass-produce toothbrushes. Prior to this, Brits were cleaning their teeth with a rag and either crushed shell or soot. Addis's toothbrushes were made from a bone handle, with horsehair, hog hair or sometimes feather bristles.

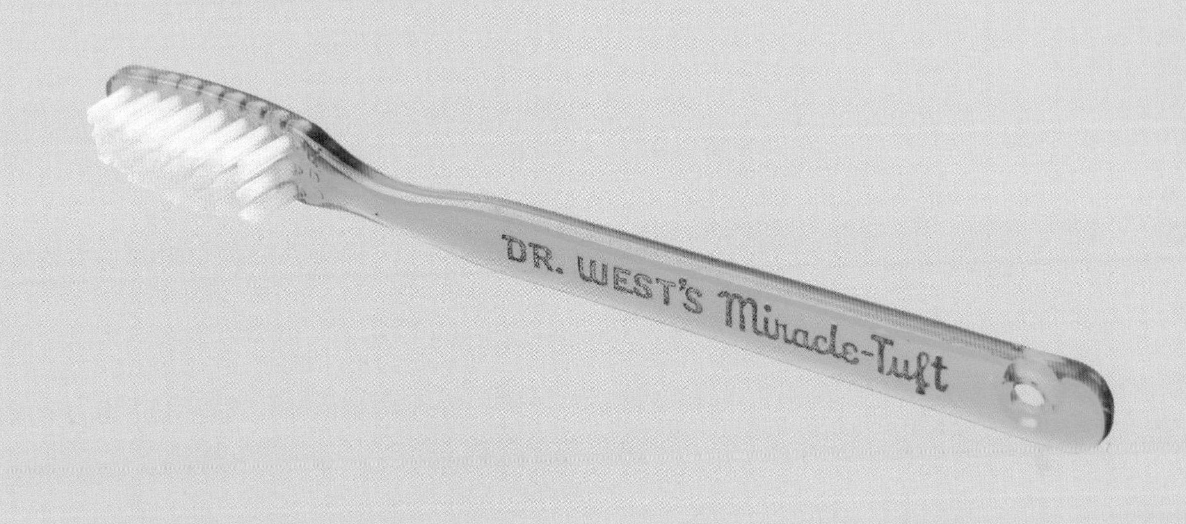

Dr. West's Miracle-Tuft toothbrush, the first to be made with nylon brushes.

Addis the company is still operating today, and has expanded into selling cleaning tools and home organising products.

In 1938, American company DuPont invented the first bristles made from nylon, making Dr. West's Miracle Toothbrush, pictured above, the first nylon-bristled and plastic-handled toothbrush. Nylon bristles are cheaper, require less effort to produce, and don't break down as easily as animal hairs, making them more hygienic and efficient overall.

The popularisation of the plastic toothbrush across the US has its roots in the military. During the American Civil War in the mid-1800s, soldiers needed to be able to rip the paper case of the bullet cartridges with their teeth when loading guns, but very few men had the six healthy opposing teeth necessary to do this. Poor dental hygiene also became a problem for the US Army when recruiting soldiers for World War I, as they had to turn away many who lacked the hardy teeth required to chew tough military rations. By World War II, the US military handed out toothbrushes to their soldiers and enforced toothbrushing in their daily hygiene routine – a habit they brought back home and normalised for the rest of the country.

Ten years after the end of World War II came the first electric toothbrush: the Boxodent, invented by Dr Phillipe Guy Woog in Switzerland in 1954. It was initially designed for people with limited motor skills, as well as orthodontic patients with braces. In general, the vibrations and motions of electric

A 19th-century toothbrush made in England.

toothbrushes help reach all areas of the teeth and gum more efficiently than a manual toothbrush. Today, different toothbrushes are continuing to be innovated to support the needs of people with disabilities. For example, the company Collis Curve makes toothbrushes for people with disabilities and their carers. They design long, curved bristles on either side of the brush head and short, straight bristles going down the middle, making it easier to reach all sides of the tooth when brushed perpendicularly. For people who may not have the mobility to stand over a sink or spit frequently, there are also toothbrushes that have liquid suction abilities at the head, with a tube attached to the end of the handle leading to a bag that collects any excess liquids in the mouth while brushing.

Electric toothbrushes have come a long way from the Boxodent, creating more effective toothbrushing for disabled and able-bodied folk alike. Unlike the electric toothbrushes of today, the Boxodent of the 1950s had a wire on the toothbrush itself, and had to be plugged into an electric outlet to be used. The system of having a charging port and batteries inside the toothbrush was introduced in the 1960s, and was safer than having a wired toothbrush. In 1992, the first ultrasonic toothbrush entered the market, using ultrasonic vibrations to break down dirt and plaque on the teeth and under the gums (technically, the ultrasonic vibrations can work without bristles). Today, we have toothbrushes built with cameras inside, behind the bristles, so you can see exactly what's on your teeth. We also have sensors that indicate where you've brushed and how effectively, and apps that connect your phone to your toothbrush for more insightful brushing. There are even U-shaped toothbrushes on the scene, which you bite onto and move side to side as the silicone bristles brush your teeth.

But toothbrushes are just half the story. Research is proving that throughout history and across the world, we've used different plants, materials and ingredient blends that have provided similar cleaning, whitening and breath-freshening results to modern toothpaste. An analysis of the skeletons from the 2,000-year-old, ancient Sudanese Al Khiday cemetery showed that fewer than 1 per cent of all the observed teeth had tooth decay – and it's believed these people farmed the kinds of grain that erode teeth. Researchers studied their tooth plaque to discover this population ate purple nutsedge, a grass-like perennial weed, potentially as food or medicine. Modern research has shown that the plant stops the growth of bacteria that commonly causes tooth decay, and it's possible this ancient civilisation came to learn this benefit through their own experience. The invention of fluoride toothpaste in 1956 marked the first time that a teeth-cleaning product was designed to prevent cavities. Fluoride works by making enamel more resistant, strengthening the teeth and reducing the risk of cavities by approximately 25 per cent.

For hygiene reasons, it's recommended to replace your toothbrush every three months, which generates a lot of waste. When placed end to end, the number of toothbrushes used in the US each year could wrap around the world three times. While a plastic handle may be more practical for drying quickly, and nylon bristles softer on the gums, this invention created a giant waste problem. There are also lots of toothbrush accessories that contribute to it – toothbrush caps, toothbrush holders and toothpaste tubes to name a few.

Interestingly, these climate concerns have led to the popularisation of bamboo-handled toothbrushes, similar to those found in the Tang dynasty all those hundreds of years ago. If the bristles aren't made with biodegradable material (such as charcoal or castor oil), the idea is to remove the nylon bristles before disposing of the bamboo handle in the compost, but this requires effort from the consumer and still leaves tiny plastic bristles polluting the environment. For now, though, this may be the most environmentally friendly option, until we find a way to ethically harvest boar hairs for the bristles – and create a major advertising campaign to convince people that rubbing boar hairs on their teeth is cool.

In 2003 the toothbrush was voted the top innovation that people couldn't live without, higher than mobile phones and cars in MIT's Lemelson Innovation Index Survey. The toothbrush, whatever form it takes, has become an integral part of our daily routine. Whether it's brushing first thing when we wake up, or after every meal, it's a habit we do almost without thinking. How we deal with a reliance on plastics for something so central to many people's dental hygiene means that the answers may rely on looking back and learning from the toothbrushes of the past.

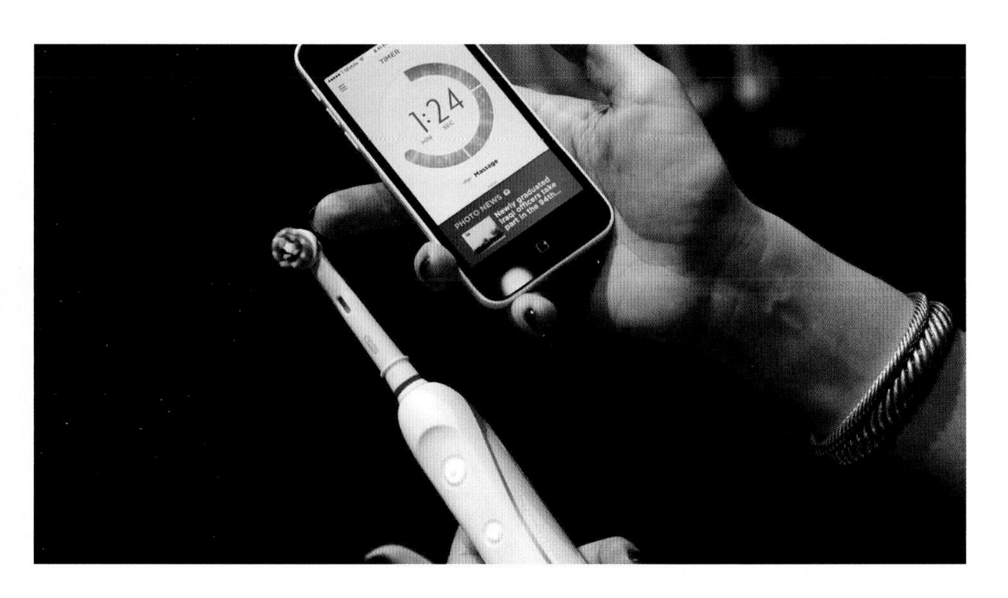

A smart toothbrush with an app to record brushing time.

KOLA NUT

Global brands don't really get much bigger than Coca-Cola. Across the world, the distinctly recognisable taste of a chilled bottle of Coke is captured by the simplicity of the company's 1904 slogan: "Delicious and Refreshing". Today, the Coca-Cola Company boasts an impressive 700,000 employees and strategically markets itself as a total beverage company, selling more than 2.2 billion servings of their drinks enjoyed in more than 200 countries and territories each day. But despite its global reach (or rather because of it) the last few years have seen Coca-Cola at loggerheads with medical authorities who have routinely linked issues such as higher obesity rates to soft drinks with a high sugar content. Given the persistence of such health warnings, it may come as a surprise to know that when Coca-Cola was first brought to the American market in the 1880s, one of its primary selling points was, oddly enough, its health benefits.

When Coca-Cola officially sprang into existence in 1886, its inventor, Atlanta-based pharmacist John Pemberton, advertised the drink as both a soda beverage *and* a nerve tonic. The latter might seem highly dubious today, but Coca-Cola was a product of its time. Post-Civil War America witnessed unprecedented levels of industrial acceleration, which in turn fostered the creation of entirely new national trade markets. During this industrial peak, American society was economically transforming at such a rapid rate that many worried that the relentless and unforgiving pace of change was causing individuals to become overworked and highly stressed. In sum, health concerns – largely over nervous disorders – characterised post-Civil War national progress, and quick on-the-go mends became all the rage.

"When Coca-Cola was first brought to the market in the 1880s, one of its primary selling points was its health benefits."

Into this social landscape Pemberton launched Coca-Cola as both a refreshing soda fountain beverage and a solution to those nervous health concerns. Coca-Cola did not radically alter the status quo at the time, since the post-Civil War beverage marketplace was already awash with all manner of cure-all soda fountain concoctions backed by multi-million-dollar advertisements. Yet, in spite of this oversaturated tonic trade, the drink did find unlikely success, and part of that success was due to one of its original ingredients – the kola nut, a largely forgotten plant product that became responsible for the "Cola" in "Coca-Cola".

Glass specimen jar containing African kola nuts, 1880–1920.

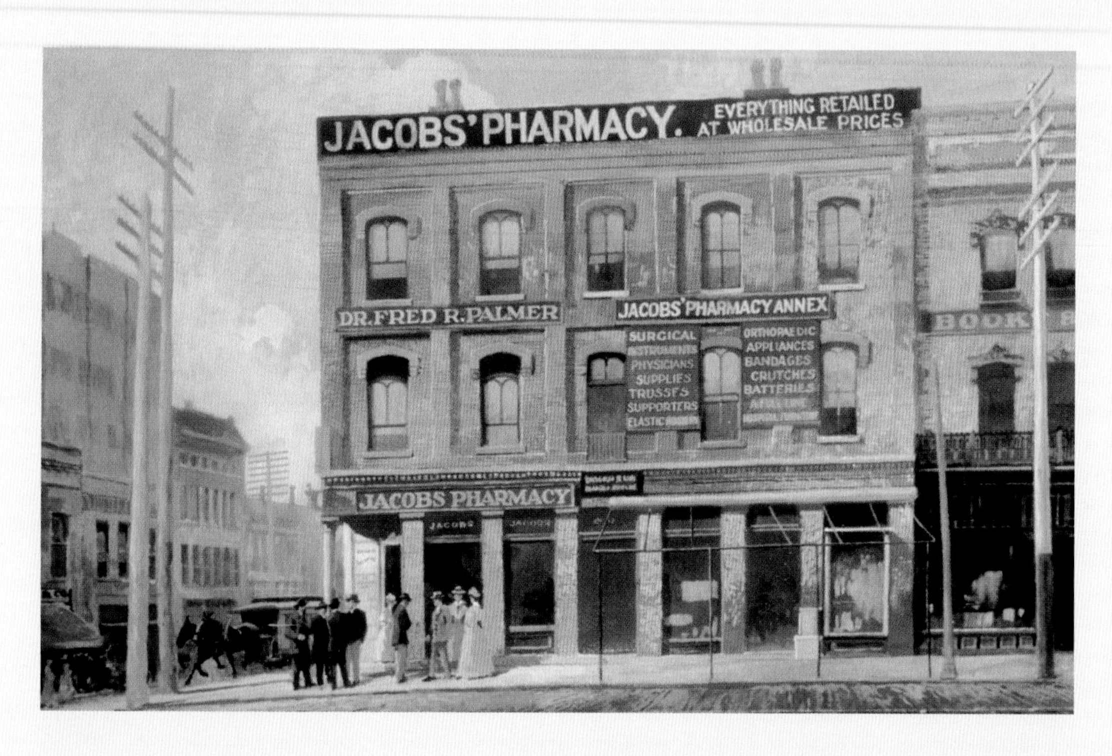

Jacobs' Pharmacy, where Coca-Cola was first served in 1886.

The kola nut, a bitter-tasting fruit about the size of a chestnut and often reddish or white in colour, is found growing on kola trees in the tropical forests of West Africa. Beyond its use for an Americanised beverage, kola nuts have for centuries served a range of religious, social, economic and health functions for West African peoples. Because of its high caffeine content, the two most common species, *Cola nitida* and *Cola acuminata*, have long been chewed by West Africans and have served as a stimulant stronger than coffee or tea. Archaeological digs in 2017 undertaken at Togu Missiri in Mali have recently revealed evidence of a thriving kola nut trade between the Western and Sahel regions of Africa dating to the 11th century, displaying the enduring use of the fruit over time.

"Kola nuts have long been chewed by West Africans and have served as a stimulant stronger than coffee or tea."

Despite the nut's wide range of uses among West African peoples, its seemingly boundless medicinal properties have historically proven most interesting to non-African peoples. According to Edmund Kobina Abaka, an expert on the history of the kola nut, its medicinal properties were first recognised outside of Africa in the 12th century by El-Ghafeky, an Arab doctor, who noted the use of kola to treat colic and stomach ache. In the late 16th century, André Álvarez of Almada then provided a more accurate description of the kola nut's medicinal properties following his visit to Guinea in West Africa. By the 19th century, following a description of the kola nut's untapped commercial potential in British publication *New Commercial Plants* in 1880,

European and American interest rapidly surged, particularly surrounding the nut's historically tantalising medicinal properties.

Thomas Christy, who authored *New Commercial Plants*, wrote in 1878 that the "indolence of the natives" was a "great bar" to the discovery and commercialisation of plants. Using remarks like these as justification, kola nuts, alongside countless other economically useful plants, were ceaselessly shipped from Africa to Europe and America for trial and testing. These trials and tests were followed by a frenzy of notes and reports in virtually all medical journals, ranging from the impressive to the fanciful. Kola nuts in particular were claimed to successfully counter seasickness, impaired digestion and diarrhoea; to allay fatigue and hunger; to sustain muscular strength; and much more. One medical journal noted that in 1888 two officers were able to scale Mount Canigou in France (just over 2,784 m high) and walk for 12 hours with only a 25-minute break fuelled simply by a small quantity of the nuts.

Kola nuts were soon advertised as a staple ingredient in the latest pharmaceutical and soda products. By 1886, Burroughs Wellcome and Company were selling their hugely popular Forced March brand, which included kola extracts, to explorers, mountaineers and travellers to offer prolonged energy. And by the early 20th century, their portable medical cases, which included packs of Kola Compound, had assisted a plethora of British Army campaigns. In America, Pemberton himself developed a medical tonic called "French Wine Coca" (the predecessor to Coca-Cola), which sold across America with considerable success. French Wine Coca, which included kola nuts and cocaine (the latter also an ingredient of the original Coca-Cola formula), was advertised in the mid-1880s as a remedy curing all those who were "suffering from any nervous complaints".

Pemberton lived to see just a fraction of the later success of Coca-Cola before dying of stomach cancer in August 1888. The Coca-Cola Company was officially incorporated in April 1888, just before his death. In 1891, American business tycoon Asa Candler purchased the rights to the company, keeping the formula under tight security. Throughout the 20th century, Coca-Cola's international fame skyrocketed and today the secrecy surrounding Coca-Cola's exact formula has assumed mythical status. But what is certain is that artificial chemicals have long since replaced the kola nut and so gone too are the drink's alleged medicinal applications.

However, despite the Western world's fluctuating interest in the nut, West Africans have never abandoned the fruit and its varied uses. In the Volta region of Ghana, it was

Men grading kola nuts, photographed in Sierra Leone, 1928.

common for teenagers in the 1960s to experiment with making their own kola drinks for fun. One particular method involved filling a clean bottle with water, cutting the kola nut into pieces, and adding those pieces to the bottle with a bit of sugar. The bottle would then be sealed with a cork and left for about a day to ferment, before it was ready to drink.

Today, a quick online search will serve up numerous videos, blogs and articles displaying the multifaceted traditional and modern uses of the nuts, from marriage ceremony symbols in the Gambia and offerings alongside prayers in Igboland to a casual snack to be shared when chatting with friends or welcoming new guests into the home.

The last two decades have seen renewed Western interest in the West African fruit. This has spurred novel multinational research developments, such as scientific studies aiming to investigate new kola varieties rich in beneficial bioactive compounds – a means of improving the quality of kola for farmers and other interested parties in countries like Ghana. But while scientific studies are currently underway, thousands of kola nut specimens remain tucked away in museum storage facilities across the UK, America and Europe, their fascinating histories of travel, trade and use largely hidden from public view.

The kola nut, an everyday West African plant product, has helped to shape global cultures of health, food, drink, religion and society in more ways than we may ever realise. Coca-Cola was an important landmark in the history of the kola nut, but the drink represents only a fraction of the kola nut's much longer life story – a story that has yet to be fully told.

With a market share of over 40 per cent, Coca-Cola is the most ubiquitous soft drink in the world, bought and sold in every country except Cuba, North Korea and Russia.

GPS

How do you get from A to B? Maybe you use satnav to plan your route, or your phone's map to find your way around. But did you know that every time you do, you're tapping into a network 20,000 km (12,425 miles) above you, which relies on atomic physics, relativity and radio waves?

The term "GPS" is commonly used to refer to any satellite navigation system, but it actually refers to the Global Positioning System created by the US government. GPS was the world's first satellite navigation system, its development beginning in the 1970s. It uses 24 satellites orbiting Earth at a height of around 20,000 km to provide full network coverage anywhere on the planet. While it was originally developed for military purposes, GPS is now used by an estimated 3 billion people worldwide, for purposes ranging from navigation and flight tracking to precision timekeeping for banking infrastructure.

But how does it work? How do the satellites know where they are, and where you are?

As you might expect, the answer is a little complicated. The system doesn't just consist of satellites – ground stations are also crucial. These are base stations situated around the world, which stay in constant communication with the satellites via radio link. Timekeeping is also of the utmost importance. Onboard each satellite is a miniature atomic clock called a "rubidium oscillator", accurate to one billionth of a second. It's not possible to service these satellites once they're launched, so the atomic clocks need to be resilient enough to withstand the harsh conditions of space, and dependable enough to keep accurate time for decades. In many ways, these atomic clocks are the unsung heroes of GPS.

"GPS uses 24 satellites orbiting Earth at around 20,000 km to provide full network coverage anywhere on the planet."

The ground stations are in fixed locations and the constant radio signals they send out enable the satellites to know where they are. The satellites themselves are constantly beaming out their own radio signals too: their exact location as given by the ground stations, and a precise timestamp thanks to the atomic clocks onboard.

GPS receivers on Earth, such as your phone or satnav, pick up these signals. These devices work out how long it's

TomTomGo 700 GPS receiver unit, 2005.

"GPS corrects any errors due to relativity, ensuring it is still accurate nearly 50 years after the launch of the first satellite."

taken for the signal to travel to them from the satellite by comparing the signal's timestamp to the current time. These radio signals travel at the speed of light. With the time taken and the speed, the devices use the equation "distance = speed × time" in order to calculate the distance that the signal has travelled; the distance between the device and the satellite.

So, now your device knows where the satellite is, and how far away it is. But how do you get from a distance to a location? Your device doesn't just calculate its distance from one satellite. It picks up signals from at least three satellites at any given time, calculating those distances too. The point where the three distances meet is the device's exact location. All this is done in a matter of seconds, and gives you your precise location to within about 5 to 10 m (16 to 33 ft).

This process is called "trilateration", and it's a one-way communication from the satellite to your device. The satellite sends out the location and time data, and your device does all the calculations to work out its location without sending anything back to the satellite. The satellites know where they are, but not where you are.

Relativity also comes into play with these calculations. According to Einstein's theories of relativity, time is distorted by both gravitational fields and relative velocity. This means that two clocks subjected to different gravitational forces will run at different rates, as will two clocks where one is moving relative to the other. This is what's happening with GPS. The further from Earth, the weaker the gravitational force, so the clocks onboard the satellites don't run at the same rate as clocks on Earth. Then, there's the relative movement due to the fact the satellites are travelling around the planet in orbit. The effects are tiny, but they add up, meaning that left unchecked they could throw off the whole system. GPS engineers made sure to account for this, and the system corrects any errors due to relativity, ensuring all timekeeping and location services delivered by GPS are still accurate nearly 50 years after the launch of the first satellite.

As of 2023, there are four satellite navigation systems in use: the US government's GPS; Russia's Global Navigation Satellite System; the European Union's Galileo; and China's BeiDou Navigation System. They all work on the same principle, and many devices actually use signals from a combination of the different systems to get the same result.

It's hard to imagine navigation nowadays without the use of these satellite systems. However, they have other applications too, such as the monitoring and regulation of power grids, the administration of banking transactions, and more. Scientists are even using GPS to alert us to natural disasters, monitor continental drift and tide patterns, and produce smart agricultural and delivery robots.

So, next time you pull out your phone to find your way to the nearest shop, or plan a route to somewhere new, think about just how much is going on in the few seconds it takes to load. And be sure to thank Einstein!

The original opening titles for *Play School*, 1964, went: "Here's a house, here's a door. Windows: 1, 2, 3, 4. Ready to knock? Turn the lock – it's *Play School!*"

PLAY SCHOOL TOYS

The toys pictured opposite – Big Ted, Poppy, Humpty, Little Ted and Jemima – are from the iconic children's programme *Play School*. Airing more than 5,000 episodes between 1964 and 1988, this pioneering series brought educational programming to pre-school children for the first time, and embodied the BBC's core values of "inform", "educate" and "entertain". *Play School* has had a lasting impact not only on the landscape of children's programmes, but on television being used as a distance learning tool for all ages. Episodes featured presenters telling stories, leading crafting activities, showing film clips of interesting places and, of course, playing with the toys.

Play School was not the first programme made for children. The BBC has a proud history of children's programming going back to the earliest days of radio, when presenters known as "Aunts" and "Uncles" read stories and sang to children. However, *Play School*'s creator, Joy Whitby, wanted the programme to be more direct than its predecessors in addressing its audience: "I was keen to address one child, not several." They were instructed to talk as if to an individual child listening at home, and not to talk down to them. Presenters often asked the audience questions, such as "Can you tell what time the clock says today?" This way of engaging the audience directly is regarded as one of the key components of *Play School*'s popularity across the decades, and has been emulated by many other programmes.

"*Play School* helped to provide both education and entertainment to children watching at home."

Play School was also special because of its target audience. It was aimed particularly at children of pre-school age, to combat what Whitby called a "dearth of nursery education". Nurseries or pre-school classes were not common in the UK at this time, so this programme helped to fill the gap and provide both education and entertainment to children and parents watching at home. As Whitby described it, *Play School* "will use all the advantages of television to do the job of a nursery school in its own exciting way."

Monica Sims, Head of Children's Programmes at BBC Television, holding the soft toy Humpty from the show *Play School*, 1968.

The approach of using television as a learning tool was adopted by the UK-based Open University, established in the 1970s to allow anyone a chance of higher education through distance learning. Among its teaching tools were programmes broadcasted publicly by the BBC between 1971 and 2006, including documentaries and lectures. This institution is now one of the largest universities in Europe, having educated more than 2.2 million people worldwide.

A typical episode of *Play School* featured a pair or trio of presenters singing songs, playing games and showing crafting activities that could be done at home. They often read stories, and guided children through telling the time. One iconic segment involved showing a short film of the outside world, taking audiences to places they might never see, like the inside of a biscuit factory, or introducing them to interesting activities such as bell-ringing. Before the film, the audience was presented with three windows – a circle, a square and an arch (a triangle was added in 1983) – and was asked which window had the film "behind" it. The camera then zoomed in on a window while the focus faded, smoothly transitioning to the pre-recorded film. Guessing the correct window was a popular game for viewers, and it was possible to determine in advance which window would be chosen. Each episode had a theme, and if that theme was, for example, wheels, then the round window would be chosen.

The programme also made history, entirely by accident. It was unintentionally the first television programme officially broadcast on BBC2. The BBC's second channel, and only the third television channel ever in the UK, was originally set to launch on the evening of 20 April 1964. However, West London experienced a power cut due to a fire at the Battersea Power Station, and the BBC's operations at Television Centre were knocked out. The BBC had to give up on their planned launch night, and shut down operations until the next morning, when they aired *Play School* at 11 a.m.

Across its lifetime, *Play School* witnessed many changes to the ways in which the UK watched television. In the early days, it was viewable only in London due to the limited range of BBC2. By the 1970s, *Play School* was viewable nationwide as the TV transmission network expanded and more people had a television set in their home. TV ownership grew from 75 per cent in the 1960s to 93 per cent by the 1970s. *Play School* also saw the switchover from black and white broadcasting to colour, and was even the first children's television programme broadcast in colour in the UK in 1968 (though it reverted to black and white for a time when more programmes were broadcast in colour and the BBC ran out of studio space).

The show had a large, rotating cast to keep it feeling fresh. At the end of the week, presenters would reference this change, saying to viewers, "Goodbye, until it's our turn to be here again." Many presenters came and went over the show's 24 years, including Virginia Stride, Gordon Rollings, Floella Benjamin and Brian Cant, to name just a few. *Play School* is credited with having the first Black presenter in children's programming, Paul Danquah, who joined in 1965.

Unlike previous programmes, where toys and puppets were carefully handled, the *Play School* toys were often played with by the presenters and so had to be replaced over the years as they wore out. The Humpty plushies in particular were prone to wearing thin and had to be replaced several times. The original Teddy was stolen in the 1970s and was replaced by the twin teddies, Big Ted and Little Ted. Hamble the doll, who was said to be the least favourite of many presenters because she was not easy to cuddle, had been made fragile by one frustrated presenter who inserted a knitting needle into the doll to keep it upright, and so was removed in 1986. The show then took the opportunity to be more representative of its audiences and replaced Hamble with Poppy, a Black doll. Incredibly, Jemima the rag doll made it from the first episode to the very last, making her one of the longest-running toys on television.

Though *Play School* has now been off air for longer than it was on, these toys are remembered by generations who grew up watching them. *Play School* transformed the landscape of programming for children and had numerous spin offs, including *Play Away*, which ran from 1971 to 1984, and *Play Days*, which ran from 1988 to 1997. Other countries – including Austria, Italy and Spain – created their own versions of *Play School* using scripts and film segments provided by the BBC, and the Australian version still runs today, making it the second longest-running children's show in the world after *Blue Peter*.

Today, we have dedicated channels for children's programming, but the legacy of a television show that proved the power of being playful and educational at the same time can be seen worldwide both on-screen and off.

BALLPOINT PENS

Ballpoint pens are the most commonly used type of pen globally, with billions made each year. What is now considered a simple and cheap office staple took the better half of a century to develop through a series of innovations, global marketing campaigns and even a patent war.

The first ballpoint pen design can be traced to American lawyer and inventor John Loud. In 1888, he patented the design of a pen that used a rolling steel ball and socket to apply ink to a surface. Loud's ballpoint pen was designed to write on rough materials like wood and leather. Unusable on paper, it failed to attract a market. Various iterations of ballpoint pens would come and go over the next 40 years, but all struggled with issues such as leaking ink, unreliable performance and fragility.

It was the Jewish-Hungarian brothers László and György Bíró who, in the 1930s, developed the first commercially viable ballpoint ink pen. László was inspired by one of the most powerful drivers of innovation: sheer frustration. As a journalist, he was tired of having to clean up the mess made by his leaky and unreliable fountain pen and its aggravatingly slow-drying Indian ink. So, he sought to create a pen that used the faster-drying inks found in newspaper printing. Teaming up with his brother György, a chemist, they devised their own oil-based ink formula, coupled with a new ball-socket design. And the first reliable ballpoint pen for use on paper was born.

The Bíró brothers patented their new design in Budapest in 1938, just before the outbreak of World War II. Forced to flee Europe in the early 1940s to avoid persecution from the Nazis, they moved to Argentina and re-established their business there. The brothers obtained a new patent in 1943 and began marketing their revolutionary pens as "Biromes". In 1943, they licensed their design to British engineer Frederick George Miles, who marketed them as "Biros". Miles produced approximately 30,000 of them for the British Royal Air Force, who found that the pen was durable and reliable at high altitudes and changing pressures.

"The British Royal Air Force found that the pen was durable and reliable at high altitudes and changing pressures."

Following the war, companies were reverse-engineering the Bíró brothers' ballpoint pen design and selling it under their own patents, resulting in a patent war across the US

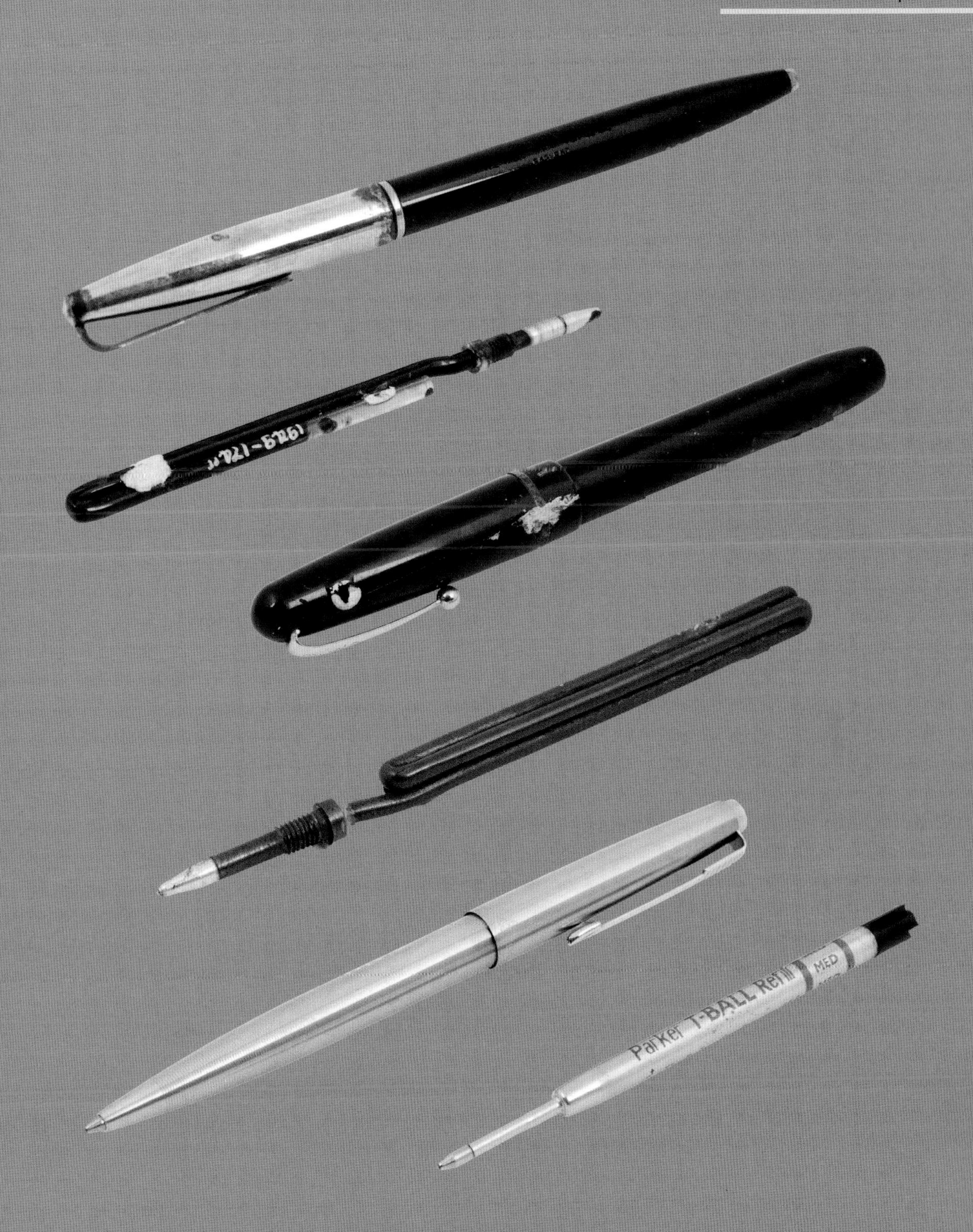

Now an affordable everyday item, ballpoint pens were expensive when first launched.

Marcel Bich, inventor of the Bic pen.

and Europe. One example was the Reynolds International Pen Company, who patented their own ballpoint pen, the "Reynolds Rocket", which became an immediate bestseller in the US. These ballpoint pens were relatively expensive and marketed as high-end, cutting-edge products. The Reynolds Rocket in 1945 cost $12.50 for a single pen (approximately $211 or £168.50 today), and UK Biros in 1950 were 55 shillings (or approximately £80 today).

This began to change in 1945, when French baron and entrepreneur Marcel Bich purchased the Bíró brothers' patent and started his own company, Société PPA (later renamed Société Bic). He worked with Swiss watchmakers to create a design that used a 1 mm ball to transmit ink without clogging or leaking. In 1950, Marcel Bich dropped the "H" from his name and began a massive marketing campaign to launch the "Bic Cristal", with reliability and an unbeatable price as its key selling points. The Bic Cristal stood out with its distinctive hexagonal design, which made it look like an everyday pencil. This contrasted with the designs of most other ballpoint pens, which imitated the look of fountain pens. By the mid-1950s, the Bic Cristal took over the European and US markets with an unbeatable price of 1 shilling or 29 US cents. By the 1960s, the design and marketing strategy of the Bic Cristal had taken the ballpoint pen from office luxury to office ubiquity.

After that, the story pretty much writes itself. The Bic Cristal remains the indomitable champion of cheap and disposable doodle devices. It is the most sold pen in history, with Société Bic celebrating their 100 billionth Bic pen sale in 2006. Its iconic design has earned the pen a permanent place in the Museum of Modern Art in New York.

"Ballpoint pens have fundamentally transformed the way we write, making the process quicker and easier. Their low cost made the act of putting pen to paper more accessible."

Despite this, ballpoint pens are still colloquially known as "biros" in countries outside of France, notably in the UK, where the Bíró brothers had some of their earliest commercial successes.

Designs of ballpoint pens have proliferated over the last 50 years. Stationery kits are a veritable smorgasbord of ballpoint varieties, including rollerballs, uniballs and gel pens, all of which share the same ball-socket design principle. Cheap ballpoint pens of different designs are commonly given away in workplaces, and companies frequently take the opportunity to brand pens with their logo. At the other end of the market, high-end ballpoint pens continue to be popular luxury products, with brands such as Montblanc, Cartier and Parker especially well known, and commemorative ballpoint pens have become popular collectors' items, such as those used in presidential bill signing ceremonies in the US.

Over 20 million Bic biros are sold each day worldwide.

Ballpoint pens have had a significant impact on our society and environment. They have fundamentally transformed the way we write, making the process quicker and easier, without having to worry about smudging or dipping your pen in an inkwell. Their low cost made the act of putting pen to paper more accessible for many, and they even have the potential to be used in extreme environments including jungles, mountains and the deep oceans. The global demand for ballpoints means more than 3 billion are made, used and discarded each year.

Unfortunately, ballpoint pens are not easily recyclable. In recent decades, improvements in recycling have made it easier to recycle the polystyrene and polypropylene bodies of these pens, but the miniscule balls, usually made of toxic heavy metals like lead and tungsten, are still very difficult for most recycling plants to process. The result is that most ballpoint pens are discarded in landfills, with billions of nearly indestructible metal balls entering the ecosystem annually.

For better or worse, ballpoint pens are one of the most important and impactful inventions of the 20th century, and a classic example of the power and problems of global mass production and consumption.

AXBT MICROPHONE

The BBC Marconi AXBT microphone is an icon of sound technologies and a symbol of the BBC that is still recognised decades after it became obsolete. These microphones were used by the BBC from the 1930s through to the late 1950s, broadcasting famous voices such as Winston Churchill and Charles de Gaulle.

The journey to the AXBT was not an easy one. When the BBC began regular radio broadcasts in the 1920s, they had to contend with the difficulties of early microphone technologies. At the BBC's first radio station in Manchester, presenters had to speak into huge funnels, directing as much sound as possible into the early microphones in order to be heard at all. The BBC decided to commission the Marconi Wireless Telegraph Company to make a microphone more suited to its rapidly growing radio service. The result was the Marconi-Sykes Magnetophone, a microphone that could pick up the sounds of insects buzzing. It was so sensitive that it had to have a special housing, made up of a rubber sling inside a huge wooden frame (affectionately known as the "meat safe", for its resemblance to a meat storage cupboard), in order to filter out unwanted vibrations. This made the whole set-up bulky and heavy.

"At the BBC's first radio station, presenters had to speak into huge funnels, directing as much sound as possible into the microphones."

Indeed, the general difficulty in moving this microphone may have been responsible for a mistake that the BBC was only recently able to own up to. This microphone was used in the BBC's famous "nightingale broadcasts" in 1924. Cellist Beatrice Harrison had noticed that when she played outside in her garden, the local nightingales joined in her music making, singing along to her playing. She convinced the director-general of the BBC, Lord John Reith, to bring a crew and broadcast this duet as the BBC's first live outside broadcast. The nightingale broadcasts captured the hearts of listeners and were repeated every year until 1942. However, in setting up the crew and equipment for that first broadcast, including this large and unwieldy microphone,

The BBC Marconi AXBT microphone, 1944–1959.

Charles de Gaulle makes one of the most important speeches in French history, considered to mark the start of the French Resistance in World War II, using the Marconi microphone in 1940.

the nightingales were scared away, and a bird impressionist had to step in at the last minute to sing along with Beatrice's cello. The BBC only confirmed this story in 2022, stating that all further nightingale broadcasts featured real nightingales.

The BBC soon moved to a much more portable and convenient microphone: the Marconi-Reisz carbon microphone. This microphone was made from a hand-sized block of marble, which was hollowed out and filled with fine carbon granules. A thin diaphragm made of mica or rice paper was stretched over the hole and, as it vibrated due to soundwaves, the diaphragm compressed the carbon granules. This created a variable electrical signal in the microphone that corresponded to the sounds it was picking up. While not quite as sensitive

"The Marconi-Reisz microphone did not require as much amplification as earlier microphones and coincided with a huge surge in the number of people using radios in the UK."

as the Magnetophone, this design was less cumbersome and did not require as much amplification as earlier microphones.

These advantages saw the Marconi-Reisz microphone being adopted not only by the BBC, but much more widely in radio and film productions throughout the late 1920s and into the 1930s. This coincided with, and probably impacted, a huge surge in the number of people using radios in the UK. In the early 1920s, around the time of the first BBC broadcasts, there were some 30,000 licences for radio receivers. By 1930, that number had risen to more than 5 million.

The range of radio broadcasts also increased rapidly during this time, both in terms of programming and how far they could reach. In 1922, broadcasts were only aired during the evenings and only available to those who lived close to the few radio stations in the UK – in London, Manchester, Birmingham and Newcastle upon Tyne. By the 1930s, the number of radio stations in the UK had increased and many more people could tune in to news, entertainment and music at various times throughout the day. Sundays were different, however, as BBC director-general John Reith only allowed broadcasts to begin after 12.30 p.m. to give listeners a chance to attend church, and the rest of the day's radio content would be solemn, focusing on religious programmes and classical music.

For the next leap forward in microphone technology, the BBC looked to the RCA Corporation in America, whose ribbon microphones were known for their superior sound quality. But when it became clear that it would be too expensive to buy and ship microphones from the US, the BBC asked their research department to collaborate with the Marconi Company to create their own ribbon microphone. The result was the now iconic Type A series, each costing less than one-tenth of the price of shipping an RCA microphone across the Atlantic.

The Type A was a huge success, with a quality of sound and reliability like no microphone the BBC had used before. This is because at the heart of each of these microphones sits an incredibly thin ribbon of aluminium, less than a thousandth of a millimetre thick. This ribbon is suspended between the two poles of a magnet and, as it is moved by vibrations in the air, the microphone converts the sounds into electrical signals. The Type A series was used for more than 20 years, and each new version was given an additional letter, leading to the fourth generation AXBT microphone, pictured on page 155.

The AXBT wasn't perfect, though. It worked well for spoken broadcasts, but it was less effective at picking up music. Additionally, it was large and heavy, and unsuitable for carrying around or operating outside. Because of this, the AXBT was mostly used in studios for dialogue. For outside broadcasting the BBC relied on more portable

Improved technology has made microphones suitable for outside broadcasting.

microphones, such as the lip microphone (also made by Marconi), which was more lightweight and could filter out unwanted sounds while preserving the voice of interviewers and sports commentators.

One iconic user of the AXBT microphone was Ahmad Kamal Sourour Effendi, also known as the "Golden Voice". Effendi delivered the first foreign language broadcast on the BBC in 1938, speaking in Arabic to listeners on the BBC Empire Service, the precursor to the BBC World Service. Effendi was already well known in the Arabic-speaking world as the founding father of the Egyptian national radio service, and brought many listeners to the BBC. This paved the way for other foreign language services in the following years. Both Effendi's voice and the AXBT microphone were known for their warm tones, making them a perfect match.

After the 1950s, further developments in microphone technologies saw the BBC move on from the AXBT. Microphones became smaller and lighter, allowing broadcasters more flexibility and mobility, and audio quality continued to improve. However, few microphones can boast the cultural impact of the Type A series. They are remembered for their longevity, their quality and their use at key historical moments, such as Charles de Gaulle's radio broadcasts to occupied France during World War II. They were such an emblem of the BBC that it was a practice to give retiring announcers gold-plated Type A microphones as a leaving present.

Today we don't tend to pay much attention to microphones. They are present in devices all around us, and we can forget how remarkable it is that we can amplify our voices to a whole stadium, talk to someone on the other side of the world, and store

"Microphones became smaller and lighter, allowing broadcasters more flexibility and mobility, and audio quality continued to improve."

music performances to be heard far into the future. But there are still huge developments happening in the audio world, as exciting new materials are used to transform how we make microphones. Graphene, which is a one-atom-thick layer of carbon atoms, is incredibly thin, strong, and flexible, making it an ideal material for the sound sensitive elements in microphones.

Spider silk is similarly thin and strong. In tests at Binghamton University in New York, it was found to improve the directional sensitivity of sounds. One application of this could be improved hearing aids that allow users to cancel out unwanted background noise and focus on what they want to hear. Professor Ron Miles of Binghamton University was inspired by how insects hear not with ears, but via small hairs on their bodies, which is what gives them such a strong sense of the direction sounds are coming from. This presents a leap forward from the days of the AXBT microphone, which was limited to only reliably picking up sounds directly in front of it or behind it, due to the limitations of a flat piece of aluminium ribbon. However, even as technology moves on, the Type A series is still a present part of our audio world, with the microphone symbol on computers and phones drawing inspiration from the design of the AXBT.

Developments in microphone technology can be seen in Britney Spears's iconic "Britney mic", allowing her to dance freely while singing.

"Puss in Boots" being shown in a "Thalia" toy theatre from the 19th century.

TOY THEATRES

Paper miniature theatre, also known as toy theatre, was a popular Regency and Victorian pastime. Originally called juvenile drama, it was first produced in England in the early 19th century and soon spread across Europe. Late Victorian toy theatres look much like the colourful German example opposite of a "Thalia" theatre, with paper characters, settings and curtains mounted on a wooden stage.

Long before television, the internet and videogames, toy theatres offered many hours of family entertainment. Young people purchased paper sheets of characters and scenes to colour, cut and mount on cardboard or wood at home. Figures could be upgraded with moving parts or attached to small sticks, wires or strings to move them more easily around the scene. When characters and sets were ready, the play could finally be performed before an audience of family and friends, either with the help of a play book (containing the script of a play) or simply using one's own imagination. Trap doors, sound effects, music and lights were frequently included in the shows.

Theatre was extremely popular in England between the 18th and 19th centuries. The Industrial Revolution caused large numbers of people to move into the big cities, especially London, creating an ever-growing audience hungry for light entertainment. However, only the upper and middle classes could afford to attend licensed theatres. The poorest crowds usually frequented unofficial performances, called "penny gaffs". These shows were also known as "blood tubs" for their tendency to perform violent murder melodramas.

This love for the theatre led to the flourishing of printing products based on contemporary theatre plays. Full-length coloured caricatures of actors and actresses from London plays, lift-the-flap children's books known as "turn-ups" or "harlequinades" (since they featured the famous pantomime character Harlequin) and booklets of spectacles, pantomimes and ballets all appeared between the end of the 18th and beginning of the 19th century.

"Long before television, toy theatres offered many hours of family entertainment."

In its original form, the juvenile drama did not resemble a full theatre in miniature. Inspired by the success of the theatre, it was launched on the market in the form of theatrical prints. Made by printing and stationery business, these prints were sold as cheap souvenirs of contemporary pantomimes and melodramas played in London theatres. Each print featured a number of characters from one play, shown in their stage costumes and striking dramatic poses. Although these products were not intended as toys, young people soon began cutting up and mounting these small portraits for display.

The earliest surviving juvenile drama print was produced in February 1811 by a London stationer named William West. The idea spread like wildfire and London publishers of miniature theatres soon increased in number. Prints were usually sold in black and white for a penny and then coloured at home, although more expensive hand-coloured versions were also available.

By 1812, several sheets of usually four characters each were produced for every play, and background scenes and front stages, copied from actual theatres, began to be made alongside them. These could be mounted on cardboard or wood to create miniature model theatres. Finally, scripts and books of words from the live plays were also printed. At this point, a complete paper play in miniature was available for boys and girls to perform all over again. Toy theatre as we know it was born.

However, many of these early plays were not full editions and presented only a small number of characters. Even fewer were accompanied by a play book. *Timour the Tartar*, published in April 1812, seems to have been the first real juvenile drama play, with several characters accompanied by different large scenes. Although it is difficult to determine whether these miniature theatres were built with the intention to perform plays, they were clearly more than just attractive drawings.

During the 1820s, juvenile drama plays soon became more sophisticated and more capable of being performed. They all presented a variety of characters, costumes and poses, together with scenes, wings and scripts. By this point, they were undoubtedly aimed at young boys with a clear taste for melodrama. Some plays, such as *Forty Thieves*, had 12 sheets of characters and 15 scenes.

Such extensive plays were not published at once, but sheets were released gradually, often in the space of a few months.

Truth be told, it was a highly competitive sector. Despite its popularity, producing theatrical prints was not a successful business. Publishers of juvenile drama plays were mostly impoverished stationers and temporary printers, who made a living out of the latest trend. The number of these companies was constantly shifting, as new businesses appeared and died within a few years.

Issuing a new play was challenging. It involved commissioning the sketches for the characters, making the printing plates and, finally, producing the prints. It all had to be done quickly, while the live play was still fresh and ideally before other publishers' editions were out.

The publisher's first move was to hire an artist who could attend the opening night at the live theatre. They made sketches for various characters and their costumes, as well as for scenes and wings. These sketches were usually transferred by the publisher into printing plates, and finally printed on paper.

The New Covent Garden Theatre, which would later become the Royal Opera House, 1810.

In the printing industry, juvenile drama prints are usually considered to be engravings even though they are technically etchings. Engraving and etching are both printing techniques involving the use of a metal plate to print images on paper. However, while engraving consists of incising a design onto a plate, etching uses an acid solution, known as the "mordant" (French for *biting*), to cut into the unprotected parts of a plate. In etching, the printing plate is coated with an acid-resistant substance usually made of beeswax, bitumen or resin, referred to as "ground".

To produce theatrical prints, the artist's sketches were usually made in lead pencil on thin paper. They could then be transferred onto a grounded printing plate by dampening the paper, laying it face down on the plate and rolling it through the press, so that the pencil lines would be forced onto the ground. They were then traced over with an etching needle, exposing the metal.

Alternatively, the drawing could be laid on top of the plate and traced over with an etching needle to make an impression on the ground underneath. Because the final print was in the drawing's reverse, these are referred to as "reverse copies".

Once the design had been traced on the ground, the plate was dipped in a bath of acid solution. The acid bit into the metal and etched wherever the needle had gone, while the rest of the plate was protected by the ground. The process could take from a couple of minutes to a few hours. Once the design had been etched into the plate, the plate was dried, cleaned from the ground, and ready for the ink. Ink was first applied all over and then wiped off, so that it only remained in the carved lines. A sheet of printing paper was then laid on top of the inked plate, a roller was passed over it, and the first print was ready.

Printing businesses were soon struggling to combine such a complex process with the increasing length of juvenile drama plays and the need to keep up with the latest live performances. At the same time, full versions of plays, consisting of tens of sheets, were getting too expensive for children to afford.

As a result, in the 1830s publishers such as John Kilby Green and the Skelt family launched a cheaper type of print. It consisted of six characters rather than four per sheet, made smaller in size and lower in quality. The style became more standardised, less accurate in copying the real theatre, and with scenes and wings often reused for many plays. This enabled a shift towards mass production, with paper theatre plays finally popularised outside London and across the UK.

More than 300 plays in miniature were reproduced by the mid-19th century. Paper theatres became the most popular toys of the time, praised for stimulating creativity and imagination in children. Many famous writers, including Charles Dickens, Lewis Carroll and Oscar Wilde, spent their childhood playing with them. In 1884, Robert Louis Stevenson even wrote an essay as a tribute to toy theatres. Its title, "A Penny Plain, Twopence Coloured", alludes to the price at which these sheets were originally sold: one penny for the plain sheets (black and white), and two pennies for the coloured ones.

Meanwhile, new printing methods were adopted. By the end of the 19th century, lithography had emerged as a much faster, cheaper and simpler method of printing. Lithography needs just one slab of stone and relies on the fact that oil and water do not mix. Designs are sketched using greasy substances or crayons onto a stone plate, so that when an oily ink is applied, it sticks only

"In the 1830s, publishers launched a cheaper type of print. This enabled a shift towards mass production, with paper theatre plays finally popularised outside London and across the UK."

to the parts of the plate with the drawing, where there is an element of grease. A piece of paper is then laid upon the picture, and the slab is sent through the press to print the image onto that paper. When no more prints are required, the stone can be cleaned and new pictures drawn. Lithography put an end to the etching and engraving trade, becoming the standard method of commercial printing in the 20th and 21st centuries.

Although lithography revolutionised printmaking, it wasn't enough to keep the toy theatre industry up and running. No new plays were published after the 1860s, as the live theatre's attention had shifted towards realism. This change in taste and the availability of new toys led to a sharp drop in popularity and by the end of the 19th century, paper theatres had become obsolete.

The old canon was kept in print by a shrinking number of publishers. Even fewer remained active well into the 20th century, such as Benjamin Pollock's Toyshop in London.

In 1877, Benjamin Pollock married Eliza Reddington, who had inherited a printing business from her father, John Reddington. A few years later, the couple decided to turn the place into a toy shop, focusing on the production of toy theatres. Many of these prints were made using John Kilby Green's printing plates, which John Reddington had purchased after the former had gone out of business.

In 1944, the shop's print collection was sold to bookseller Alan Keen, who modernised the stock for a contemporary audience and collaborated with famous artists and performers. Actor Laurence Olivier even contributed to a toy theatre version of the 1948 film *Hamlet*. The business changed ownership again in the 1950s, and the stock expanded with the acquisition of Skelt's printing plates.

Fascination for the miniature theatre made a comeback in the 1980s, when artists all over the world began revisiting the genre, either using old paper examples or adapting them to produce modern performances. In the 1990s, Benjamin Pollock's Toyshop was bought by brothers Christopher and Peter Baldwin. The latter was a toy theatre collector and actor best known for his role in the British soap opera *Coronation Street*.

The shop, situated in Covent Garden Piazza, is now run by Louise Heard. It sells reproduction and original toy theatres from around the world, as well as its own range of paper models designed by contemporary artists. These creations are often displayed at top London locations like the Royal Opera House.

It was Peter Baldwin who selected for the Science Museum the Thalia theatre pictured on page 160, and as he once said of his toy theatre fascination: "You can create magic with it by changing the scenery, moving characters on and off, by raising the curtain, changing the lighting. You become manager, actor, producer, everything." More than just toys or a pastime for boys, paper theatres have ignited children and adults' imagination for over two centuries. Thanks to shops like Pollock's, these nostalgic toys may continue to spark dreams for years to come.

LOCKS

It's rare that any of us are actively invited to pick a lock. Yet this is what Bramah & Co challenged passers-by to do with this padlock at their shop in London in 1790. Flaunting their reputation for creating the most secure locks, they displayed the "Challenge Lock" opposite in the window, accompanied by a board that read: "The Artist who can make an instrument that will pick or open this lock shall receive 200 guineas the moment it is produced."

The challenge stood for an impressive 67 years, during which time Joseph Bramah died, leaving the company to his sons. During the Great Exhibition of 1851, lock-picking competitions between rival locksmiths began captivating the public, and Bramah's prestigious lock caught the attention of American locksmith Alfred Charles Hobbs. Hobbs' success in picking it launched what became known as the Great Lock Controversy, a period of public fascination and debate about what this meant for security in Britain and beyond.

The oldest known lock, estimated to be 4,000 years old, was found in the Khorsabad palace ruins in modern day Iraq. It is an early version of the type of lock known today as the "pin tumbler", which uses wooden pins of varying lengths to prevent a door from opening. An inserted key pushes up the pins, allowing the wooden bolt to move across and unlock the door. Although simple, this design would form the basis of the lock patented by American lock inventor Linus Yale Sr. in 1843.

Metal padlocks are thought to be Roman in origin, with multiple examples discovered at sites such as Pompeii. In the Middle Ages, Scandinavian civilisations were particularly skilled at making padlocks, and several were found in York during excavations of a Viking settlement there. Medieval art, such as carvings, illuminated manuscripts and stained-glass windows, illustrate the intricacy of the locks produced in this period, especially in monasteries and churches.

"During the Great Exhibition of 1851, lock-picking competitions began captivating the public."

The Beddington Lock illustrates the importance of locks to royalty in the Tudor period. It is an elaborate gilded lock with a central sliding plate concealing two key holes that is decorated with the arms of Henry VIII. In 1539 it

Bramah padlock and key, 1801 – the challenge lock of the Lock Controversy of 1851.

was extracted from a door at Beddington House, the seat of the Carew family, when it was seized by Henry VIII after Sir Nicholas Carew's execution. It then accompanied Henry VIII as he travelled around England, and was installed on his chamber door wherever he stayed to ensure his security.

By 1800, rapidly growing towns and cities became notorious as centres of criminal activity. Numbers of highly skilled burglars who could pick locks were increasing, and a growing middle class valued their property and wanted to protect it, driving a boom in patents for locks. In 1817, a burglary occurred in Portsmouth Dockyard using false keys. The government responded by announcing a competition to create a lock that could be opened with only one unique key. Jeremiah Chubb was working as an ironmonger in Portsmouth and responded by inventing his "Detector Lock".

Chubb developed a four-lever lock that would stop working if someone attempted to pick it, and required a specific key to reset it. He wrote in his patent: "In this state the lock is what I call detected and the possessor of the true key has evidence that an attempt has been made to violate the lock, because the true key will not now open it."

The government promised free pardon and £100, worth over £7,000 today, to a locksmith who was a convict aboard one of the prison ships in Portsmouth Docks if he could successfully pick Chubb's lock. The convict had successfully picked every lock given to him and was confident, but after three months he admitted defeat.

Chubb's locks became famous and even made an appearance in two of Arthur Conan Doyle's stories about his fictional detective Sherlock Holmes. In *A Scandal in Bohemia* (1891), a house is described as having a "Chubb lock to the door", and in *The Adventure of the Golden Pince-Nez* (1904), the difficulty of picking a Chubb lock is a clue in the plot.

One advertisement even adapted Percy Bysshe Shelley's poem "Ozymandias" (1818) for comic and promotional effect: "My name is Chubb, that makes the Patent Locks; Look on my works, ye burglars, and despair!"

In 1823, Chubb became the sole locks supplier for England's post offices and prison service. Chubb's company was even commissioned to create a secure display case for the famous Koh-i-Noor diamond, then the world's largest and in British hands following the annexation of the Punjab at the end of the Second Anglo-Sikh War.

The Detector Lock remained apparently infallible until 1851, when Alfred Charles Hobbs picked it in just 25 minutes at the Great Exhibition in front of a group of fascinated onlookers, before taking on Joseph Bramah's Challenge Lock.

By breaking both the Chubb and Bramah locks, Hobbs simultaneously became the most celebrated lock picker of his day and created panic among British bankers who relied on these locks. This was precisely the effect he had been creating through a string of lock-picking displays in America, conducted on behalf of the lock company Day and Newell to expose the vulnerability of their rivals.

Hobbs' playful delight in his lock-picking proficiency is suggested in a book written by Reverend Samuel Orcutt. In 1848, Hobbs responded to an advertisement by a Mr Woodbridge of New Jersey, who offered $500 to anyone who could break the safe in the Merchant's Exchange reading room. This lock had 479,001,600 possible arrangements of the pins inside, and was engineered so that picking tools would get stuck inside the lock.

After working on the lock for a few hours one evening, Hobbs returned the next morning to find a gathered crowd. He called for Woodbridge, who asked what the trouble was.

"There's something the matter with the lock," said Hobbs.

"What is it?" replied Woodbridge.

"Your lock won't keep the door shut," Hobbs declared, opening the safe door.

It is little surprise then that Hobbs had the famously "unbreakable" Chubb and Bramah locks in his sights when he travelled to London for the Great Exhibition. Having packed six drawers of lock picking tools into a trunk, he had to obtain a letter from New York City's chief of police testifying his good character to get him through customs.

Challenging the two most famous British lock makers at the biggest showcase of British imperial and industrial power ever seen was audacious, even for Hobbs. In doing so, he would expose a flaw in the security system at the heart of the British Empire and leave the press and public reeling.

This sense of disbelief fuelled the Great Lock Controversy, in which the British public and press debated the truth of what had happened. Some were sceptical about whether it was all a façade, arguing that Hobbs had cheated.

Meanwhile, British lock companies and the British establishment tried to downplay Hobbs' achievement. American publication *The Bankers' Magazine* wrote that "the result of the experiment has simply shown that, under a combination of the most favourable circumstances, and such as practically could never exist, Mr Hobbs has opened the lock."

But Britain's reputation as the world's best lock maker was already tarnished. The Bank of England itself replaced Chubb's locks for those of Day and Newell. Long term, however, both the Chubb and Bramah companies continued operating successfully with new improvements inspired by Hobbs' picking. Both companies still exist today. Hobbs stayed in London for nine years and began his own lock company, until 1954 when it was bought by the Chubb Company.

Today, security can feel far less reliable, with digital data breaches and hacking a regular occurrence. For the Victorians, however, the experience of discovering the vulnerability of supposedly impenetrable locks must have been deeply shocking. A prominent locksmith, James Tildesley, observed in the 1860s that the Great Lock Controversy "gave a stimulus to the lock trade, such as it has never received before or since." *The Spectator* noted that "Hobbs dispelled the illusion and set the lock-making trade free." Hobbs fractured the untouchable image of British lock companies and in doing so catalysed technological innovation in security which continues today.

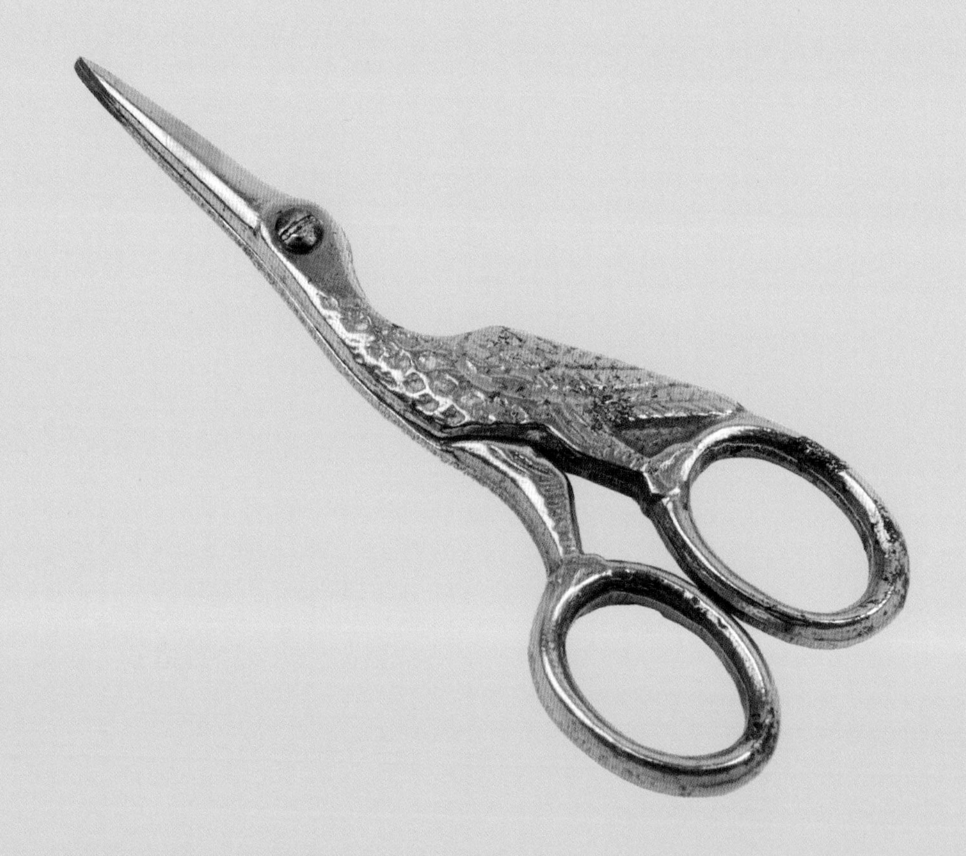

Presentation scissors used for the launch of the Great Eastern Railway twin-screw steamship *Vienna*, Hull, 18 July 1894.

SCISSORS

At first glance you may not expect the artist Henri Matisse to have anything in common with both a hairdresser and a surgeon. But looking at their tools, you will see that scissors are vital to the work of each: Matisse created colourful paper cut-outs (which he called "drawing with scissors"), while hairdressers and surgeons have multiple specialist pairs. As well as professional uses, scissors are used in daily life to shorten string, trim nails and open parcels, making them a common tool in many households. As something that has existed for thousands of years across the globe, scissors have also taken on a significant amount of spiritual symbolism in that time.

The history of scissors contains a surprising amount of mystery. For example, anyone familiar with a sewing kit will likely recognise the popular stork-shaped design of embroidery scissors, like the pair pictured opposite. There is a myth that they were used both for embroidery and for cutting the umbilical cord after birth, due to the stork's association with babies. However, scissors used for such a purpose also feature clamps in the design for restricting the blood flow of the umbilical cord, rather than severing it. Scissor specialists such as William Whiteley believe that midwives from the 1700s used their sewing kits to pass the time during labour: this association led to the embroidery scissors taking on a similar design to the clamps. The pair in the Science Museum's collection (measuring only 4 cm, or less than 2 inches, long!) is from the 19th century, meaning that regardless of the reasoning behind the design, it has had an enduring popularity for over 100 years.

"Spring scissors – two blades held together at the end with a flexible strip of metal — existed in Mesopotamia at least 3,000 years ago."

Another myth is that scissors were invented by Leonardo da Vinci. While he was certainly a successful inventor, and even attempted early designs of "flying machines", scissors cannot be credited to him. Spring scissors, which consisted of two blades held together at the end with a flexible strip of metal, existed in Mesopotamia at least 3,000 years ago. The blades were squeezed together to cut, and the strip of metal sprang them open again when the hands released the pressure. Similar designs were created afterwards in Egypt, and later by Rome during the Roman Empire and China during the Han dynasty. Pivot-style scissors can be traced back to the Romans of 100 CE. These are the type of scissors you probably have in your home: two blades are

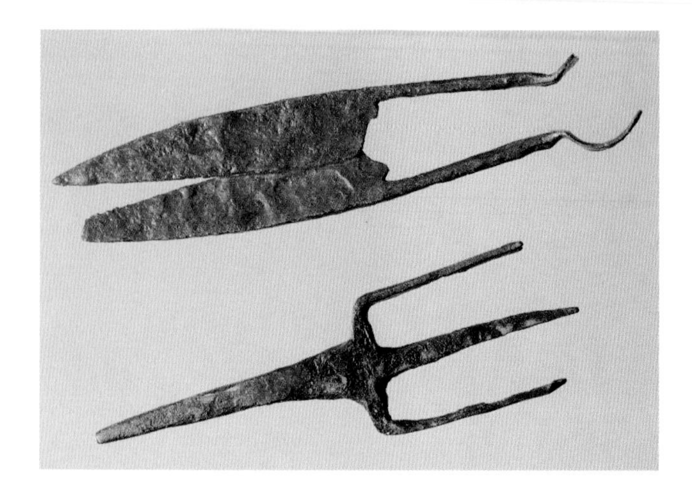

Ancient Gallo-Roman scissors.

held together in the middle at a pivot point, or a "fulcrum". The handles at the end are opened and closed by hand, with much more control than spring scissors. Throughout history, the materials used to make scissors have changed alongside the development of metals, moving from bronze to iron to modern-day steel.

The majority of scissors today are produced in the same way and sold at affordable price points. Manufacturers use machines to rapidly punch the shape of the blades – called "blanks" – out of large sheets of metal. Scissors were first mass-produced in Sheffield in 1761, the home of British steel production. Traditional scissor companies soon adopted the same manufacturing process, such as

"There are many beliefs, rituals and superstitions related to scissors across the globe. Scissors can be used for spiritual protection."

Fiskars in Finland, known for creating the first plastic-handled scissors, and Hangzhou Zhang Xiaoquan in China, which had already been making scissors since 1663. As Chinese household scissors are multifunctional and often used in cooking, they tend to have shorter blades than Western scissors, which are most commonly used for cutting paper. As a result, Hangzhou Zhang Xiaoquan has also branched out to global markets by producing pairs with longer blades.

Traditional, hand-forged scissor-making is regarded as a critically endangered craft by the Heritage Crafts Association, though there are still those who take pride in incorporating traditional methods. Sheffield's respected scissor manufacturers, William Whiteley and Ernest Wright, employ some of England's only traditional scissors grinders and "putter-togetherers" – the official title of a fully trained scissors craftsperson. In the same city, metalworker Grace Horne works in a slightly different factory: a converted Victorian public toilet! She specialises in bespoke, collectible scissors, often crafted with intricate detail and patterns. In Japan, blacksmith Yasuhiro Hirakawa uses katana-forging techniques to create custom pairs, sometimes taking up to a year to finish. He has been making scissors for more than 50 years, including the most expensive pair of bonsai scissors ever, worth 3.6 million yen (around US$24,000, or £19,000).

Despite the rich and lengthy history of scissor-making, the first successful left-handed scissors were not created until 1967. To count as left-handed, the blade that moves upwards when cutting must be on the left-hand side. Previous attempts did not function properly because they simply involved reversing the handles. When using the correct pair, the thumb pushes downwards and to the side, which brings the blades close together for a clean cut; when using the wrong pair, it pushes them apart. Additionally, using the wrong pair means that the top blade blocks the user's view, hindering accuracy. There have been attempts to make ambidextrous pairs, but as with early left-handed scissors, most designs only properly cater to use by one hand, as the top blade remains on the same side even when turned over.

Whether due to the material used, the sharpness of the blade or the symbol of a cross that an open pair of scissors resembles, there are many beliefs, rituals and superstitions related to scissors across the globe. The popularity and depth of these beliefs varies over time and even within specific locations. One traditional belief found in numerous countries, including England and Jamaica, is that giving scissors as a gift will "cut" a friendship. Therefore, a small coin should be given in return to "purchase" them. This tradition dates back hundreds of years; in the Tudor period, Queen Mary I had a specific collection of coins that she received in return for gifts of knives and scissors.

Scissors can also be used for spiritual protection. Folktales from Canada describe placing a pair by windows to prevent ghosts from entering, while people in England previously hung them open in a "cross" shape above doors to ward off witches. A knife or scissors may also be placed on the chest of a person after death. In Denmark, this prevents the body from reanimating, and in Japan it protects the body from spirits.

Some actions have differing interpretations depending on the culture. In Egypt, for

Dress-maker's shears have long, tapered blades sharper than typical scissors.

instance, some believe you should never open scissors without using them to cut, while in Mexico, an open pair may be placed under the bed to aid sleep. In China, people may avoid using scissors during pregnancy to ensure the health of the baby, whereas in Iran, a pregnant person may choose between a knife and scissors while blindfolded to predict the sex of their baby.

As a result, scissors can come to symbolise a protective token or a curse. Perhaps it isn't surprising that such a variety of meanings have been attached to scissors. After all, they have been used in everyday kitchens and workshops for thousands of years. The variety of uses for scissors may only be outnumbered by the amount of beliefs surrounding them.

PREGNANCY TESTS

Before the Covid-19 pandemic brought lateral flow technology into all our lives, pregnancy tests were among the most recognisable and widely used home diagnostic tests. With generic strips now costing pennies, it's easy to take for granted the ability to find out in minutes if you're pregnant within the comfort and privacy of your own (or indeed any) bathroom.

For much of human history, the process was far less straightforward. Until as recently as the 1960s, most people relied on observable changes in their bodies for diagnosis, seeking medical advice only once a pregnancy was well established. But early signs and symptoms can vary between individuals, making it tricky to be certain before the bump appears and the first foetal movements can be felt.

Attempts to develop a pregnancy test date back at least as far as 1350 BCE. Ancient Egyptian texts from this time suggest urinating on wheat and barley seeds – the idea being that if they sprouted, you were expecting. Remarkably, when researchers replicated the test in 1963 they found it had more than a grain of truth, correctly identifying 70 per cent of pregnancies. While ancient medics believed some kind of vital force exuded from the body during pregnancy made plants (as well as people) grow, modern scientists have attributed the effect to elevated oestrogen levels in pregnancy urine.

By the 11th century, uroscopy – scrutinising a patient's urine for clues – had overtaken peeing on cereals as the diagnostic method of choice. Persian polymath Ibn Sina (commonly known in the West as Avicenna) described the urine of someone in early pregnancy as pale yellow and cloudy, "as if there is cotton scattered in the middle of it". In the centuries that followed, so-called "piss prophets" used sight, smell and even taste to try to identify signs specific to the urine excreted during pregnancy. With such techniques having dubious efficacy, it's likely these medieval physicians drew their

"'Piss prophets' used sight, smell and even taste to try to identify signs specific to the urine excreted during pregnancy."

"Clearblue One Step" home pregnancy test kit with urine sampler giving result in three minutes, England, 1988.

conclusions more from the careful observation of their patient rather than from the pee.

Yet their efforts weren't entirely misplaced: today's pregnancy tests also detect changes in urine. The first reliable biomedical test was developed in 1927 by German gynaecologists Selmar Aschheim and Bernhard Zondek, after a hormone called human chorionic gonadotropin (hCG) was discovered in the urine produced following recent conception. Produced by the growing embryo, hCG plays an important role in maintaining the pregnancy during the first trimester. What's more, using methods no less bizarre than their predecessors, Aschheim and Zondek found that the presence of this hormone could be confirmed by certain animals.

The pair realised that, when injected with pregnancy urine, immature female mice would go into heat, causing their ovaries to grow. Unfortunately, this reaction could be ascertained only through dissection. Later versions used rabbits, and for a time "the rabbit died" was another way of saying someone was pregnant. They were highly accurate, but five animals were needed per test, making it expensive and resource-heavy.

In the mid-1930s, scientists began looking for a more sustainable model; an animal that would produce observable changes in response to hCG without needing to be dissected. Bitterlings, a carp-like fish, seemed to provide the answer. Females of this species have an ovipositor or tubular organ for laying eggs, which was shown to expand within hours when human pregnancy urine was added to their water. Following the test, the fish could be returned to a tank of fresh water and reused after three weeks.

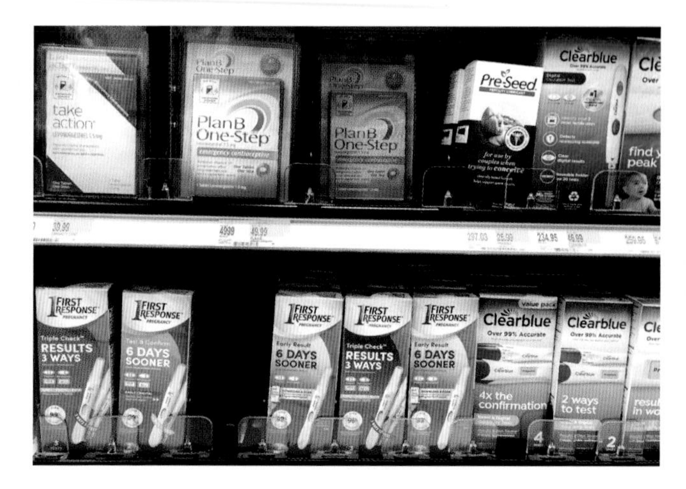

Modern pregnancy tests are more affordable, accurate and widely available than their early predecessors, allowing discovering pregnancy early to be easier and more personal than ever before.

Hearing of the discovery, an enthused *TIME* reader predicted that "every standard American home" would soon be "equipped with an aquarium containing a female bitterling that would be as handy as the radio, the vacuum cleaner, the bottle of antiseptic, etc. ... especially in times when the budget of most households does not permit haphazard payments to obstetricians!"

The hype proved short-lived, however. Further studies found urine samples from postmenopausal women – and even men – produced positive results, rendering the bitterling test about as useful as uroscopy for diagnosing pregnancy.

Meanwhile, researchers in South Africa had come up with a new solution: the Xenopus frog. Like mice and rabbits, female frogs of this species respond visibly to injections of urine containing hCG, in this case by spontaneously laying eggs. Unlike their furry predecessors, the amphibians survived the experience and could be retested time and again over their 30-year lifespan. Results were nearly 100 per cent accurate and much quicker than the "rabbit test", being obtained in hours rather than days. Frogs went on to become the standard means of diagnosing pregnancy right up until the 1960s, when immunological tests that could detect hCG without the need for live animals were developed.

But despite thousands of unsuspecting creatures being jabbed over this period, pregnancy testing was far from routine. The animal-based tests were considerably less sensitive than today's sticks, so had to be taken at least a month after conception and were largely reserved for problematic cases, such as unusual bleeding or suspected cancer. Doctors therefore generally advised an approach of "wait and see".

The turning point came in the 1970s, when a heightened desire to detect pregnancy as early as possible – a need influenced by advances in prenatal care and liberalised access to abortion in North America and Europe – coincided with new research. It was during this decade that home pregnancy test kits became available for the first time, finally removing medical professionals from the moment of truth.

The first of these, Predictor, was the brainchild of American graphic designer Margaret Crane. Crane was freelancing at the pharmaceutical company Organon when she noticed that the new immunological tests (until that point only accessible to consumers directly via commercial laboratories) were relatively straightforward to interpret. Taking inspiration from her desk tidy, she designed a prototype with all the components needed to perform the test at home.

The process was fiddly, took around two hours and, while cheaper than lab tests, remained costly. When Predictor was launched in the UK in 1971, it retailed at £1.75, which is about £30 in today's money. Yet this "private little revolution" (as a 1978 advert for a rival home pregnancy test put it) offered not only the ability to make this potentially life-changing discovery at a convenient time and place, but also offered the power to make decisions without unwanted judgement or coercion from the medical establishment.

By the late 1980s, lateral flow testing had eliminated the need for miniature chemistry kits and the "pee-on-a-stick" method was born. Today's early pregnancy tests are sensitive enough to detect hCG from as little as ten days after conception. The wait may have become a lot shorter, and the process less mediated, but we owe a debt to the ancient Egyptians who first looked to urine to determine if someone was pregnant.

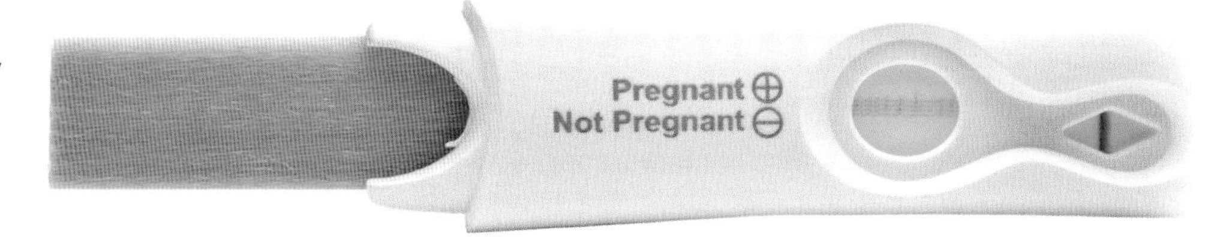

A home pregnancy kit displaying a negative result.

FIRST AID KITS

For minor injuries or pain relief, many of us will reach for our first aid kits, if we have them. Stored in a variety of containers, boxes and bags, and easy to find in an emergency, these kits can be as individual as the people that own them. Examining their contents reveals much about their owners' attitudes to what they could and can treat at home and what medical system of thought they use to treat themselves.

Designed to be used by the person who is ill or injured, or by the people who are with them, first aid kids may contain a mixture of medical treatments from across the globe, as well as home remedies. The UK National Health Service recommends that first aid kits contain plasters, bandages, tweezers, scissors, painkillers, antiseptic creams, eye wash, thermometers and relief for insect bites and stings. Traditional Chinese medicine might advise powders, balms and ointments to stop bleeding, treat burns and offer pain relief for bruises. Ayurvedic medicine, a holistic medical system based on ancient Indian writings, may recommend using ingredients already found in kitchen cupboards, such as honey and turmeric for a sore throat, or a paste of dry ginger for a headache. Some first aid kits remain at home, while others are adapted for workplaces, schools or travel, including for expeditions.

In the 1870s, German surgeon Friedrich von Esmarch is believed to have coined the term *erste hilfe*, or first aid. Esmarch trained soldiers to administer help to troops on the battlefield, with bandaging and splinting being the two main ways they could save someone's life. Introduced in 1855, British soldiers carried versions of field dressings – bandages and safety pins that came with instructions for use. Larger and more specialised first aid kits were used at different points along a wounded person's journey away from the front lines, through the chain of evacuation developed during World War I. First aid training was introduced during World War II for all British Army personnel, and it continues today.

"German surgeon Friedrich von Esmarch is believed to have coined the term *erste hilfe*, or first aid."

While the battlefield and naval services have seen necessary advances in first aid, people have always found ways to treat themselves, using plants from the natural world, or providing a place of care and comfort. Armed forces and

"Tabloid" first aid kit, used on the first direct transatlantic flight by airship, 1919.

A pharmacists' portable apothecary.

sailing ships used some of the earliest medicine chests and containers, dating to at least the 1st century CE. In the 1500s and 1600s, medicine chests for those who could afford them contained ready-made medicines for fevers, wound treatments and ingredients to make others. A set of scales and weights, a pestle and mortar, and a funnel to refill bottles often sat alongside ingredients. These medicines were all based on the European medical model of the 1500s and 1600s, which posited that the body was made up of four humours that needed to remain in balance with each other for a person to remain healthy. Treatments to remove an excess of a humour included medicine to induce vomiting or to be used as enemas. Liquid painkillers were always based on opium, now known to be dangerously addictive.

From the 1770s onwards in Britain, families could purchase medicine chests as a first line of defence for their healthcare needs. Chests contained powdered and liquid treatments for pain, fevers, wounds, diarrhoea, burns and stimulants. All medical care had to be paid for at this time, so physicians carrying out house calls would also use these home medicine chests or families could treat themselves if they had the skills to do so. By 1900, these medicine chests were going out of fashion as new forms of medicines became available.

Compressed tablets were brought to the UK by two Americans, Silas Burroughs and Henry Wellcome. Called the "Tabloid" brand, these were tiny, compressed tablets with accurate doses, and could be transported far more easily than a medicine chest fitted out with heavy glass bottles and wooden cases. Burroughs Wellcome & Company prototyped their first portable medicine case after a chance meeting. In 1881, Silas Burroughs met Dr Colin Valentine, founder of Agra Medical Missionary Training Institute, who suggested that a small case fitted out with compressed medicines could be of benefit to medical practitioners and missionaries in India. In return for trialling Burroughs' prototype first aid kit and writing an accompanying medical guide, Valentine received a donation to his Institute. It is unclear how Valentine used the prototype, whether he used it to treat people he encountered during his missionary work, or whether he self-administered the contents or if he added to it using local knowledge.

Encouraged by their initial success, Burroughs Wellcome & Company realised the potential and expanded into making a range of first aid supplies and medicine chests, tailored to the needs of the user and packed into specially designed cases. The company gave them to those travelling as part of the British Empire's colonial expansion in Africa.

By exporting their medicines throughout British colonies and territories Burroughs Wellcome & Company were also exporting their ideas about health, science and medicine, in line with white European ideas. The colonial slant of their branding is clear in the company's publication, *The Romance of Exploration and Emergency First Aid from Stanley to Byrd*, in which they state: "The medicine chest goes hand in hand with the advance of civilisation." This was stated with little recourse to recognising and valuing local culture and systems of medical thinking and treatment. Often these kits contained medicines using plants from all over the world, repackaged into tiny, compressed tablets, with scientific studies backing their efficacy.

First aid kits and medicine chests were given to Ernest Shackleton during his attempt to reach the South Pole between 1907 and 1909. British teams attempting to reach the summit of Everest in the 1930s also received them. The first ever transatlantic balloon flight by Walter Wellman in 1910, and the first ever non-stop transatlantic flight by John Alcock and Arthur Brown in 1919, also carried Burroughs Wellcome & Company kits. Alcock remarked that their first aid case "is the only possible medical outfit for an aviator". Once expeditions were over, many explorers donated their kits back to Burroughs Wellcome & Company to be used as advertising material, particularly at the Chicago Exposition of 1934. They later found a home in Sir Henry Wellcome's museum collection, a burgeoning attempt to collect the history of humans' lived experience with a particular focus on their health. Not only were the Burroughs Wellcome & Company kits linked with "civilisation" but that of progress, scientific endeavour and reaching the limits of human potential at any cost.

Burroughs Wellcome & Company kits went through several stages of prototyping and development. Up to the 1930s, their product line rapidly expanded, with kits for home, work, car travel, yachting, Boy Scouts, theatres, air raids and gas attacks, and even snake bites. In the 1920s, their "Pac-Kit" was developed when a telephone company's insurance arm asked for a special kit for their workers to treat minor injuries while at work. Single dose packets of smelling salts and iodine for cleaning wounds were included, giving the kit its name. "Pac-Kit" was also a play on words, enforcing the importance of "packing it" with you when on the go.

But another company might have a claim to selling the first portable first aid kit. According to Johnson & Johnson's company history, the first ever first aid kit was created in 1888 following a chance conversation between Robert Johnson, co-founder of the company, and a railway surgeon. Many railway accidents happened in remote rural areas, far from the reaches of medical services, and the often dangerous nature of the work necessitated portable first aid supplies. A year later, the company produced more than 18 different types.

Many manufacturers, including Johnson & Johnson and Burroughs Wellcome & Company, placed a red cross on their early first aid kits, in violation of the symbol's use. The symbol was adopted by the International Committee of the Red Cross, the Geneva-based humanitarian organisation, in 1864 to designate medical services, protecting those who are wearing it by international law. Its use is still closely regulated. Since the 1930s, a white cross on a green background has become the internationally recognised symbol used to denote first aid kits and stations, and places of safety or help. These symbols are often found in workplaces, public buildings and transport hubs.

First aid kits may contain everything you need to treat minor injuries or ailments, but without the knowledge to use them, they are ineffective. First aid training has often been provided by volunteer organisations such as St John Ambulance and the Red Cross.

A standard first aid kit in Britain must contain plasters, sterile dressings, cleansing wipes, tape, disposable gloves, a resuscitation face shield, a foil blanket and a burn dressing, as well as clothing cutters, disposable gloves and usage instructions.

Booklets, videos and manuals have all been used to support such training. In the 1910s, cigarette manufacturers WD & HO Wills produced a series of 50 cards forming a how-to guide for treating broken bones, applying tourniquets, bandaging and even resuscitation methods. While we are now aware of the dangers of tobacco smoking, it is estimated that in 1950 up to 80 per cent of adult men in the UK were smokers, so there was a huge market for cigarettes. Originally cigarette cards bolstered the packaging but from the 1900s to the 1950s they became used for advertising or imparting knowledge with a large potential audience, and were highly collectible. More recently, first aid training can be found on smartphone apps.

First aid is extending beyond the physical. Mental health first aid kits are something anyone can make. Each is made by and tailored to an individual, being filled with items to comfort and calm people and provide coping strategies. While not available to buy as a kit, charities such as ChildLine and some NHS Trusts recommend including a list of local services and websites, any prescribed medication, and ways to distract yourself, such as exercises, relaxation and sleep tips, comfortable clothes, music, stress balls, crafting kits, and ways to write, draw or express your feelings. Kits can be updated and adapted to different circumstances, for instance being at home or travelling. Some people may prefer an entirely digital version with apps and calming techniques on their smartphones. Others may prefer a combination of the two. Just like a kit for cuts and bruises, a mental health kit should be there when someone needs it, filled with what they need in that moment.

TEACUP

Chances are, you probably own at least one teacup or mug. Maybe you're drinking from one right now, an everyday favourite, perhaps, much-loved but slightly chipped and scratched. Or something a bit fancier, like the charming pale-blue teacup on page 184, from the 18th century with painted flowers and gilt decoration. Whatever you have, it's likely that it's made from some kind of porcelain, like bone china, or "hard-paste" porcelain, as this one is.

Today, cups, mugs and all sorts of tableware in these ceramic materials are plentiful, from the cheap and cheerful to the considerably more costly. So it's hard to imagine that porcelain, specifically hard-paste porcelain, was once exclusively a luxury good in Europe, referred to as "white gold". That's because the scientific secrets of its chemical recipe and technological manufacture were known only in China, where hard-paste porcelain has been made since the 6th century CE. It would be more than a thousand years before Europeans cracked the code in the early decades of the 18th century.

Until then, they had to rely on expensive imports via Portuguese and then Dutch trade with China, or have a go at making their own imitations. But nothing could quite match the qualities of the real deal: brilliantly white, beautifully translucent, delicate to the touch, yet also (as its name suggests) incredibly strong and resistant to heat and wear. In fact, porcelain became so synonymous with China that we still tend to refer to goods made from this material as "china".

His name may not be as familiar in the world of ceramics today as that of, say, Wedgwood, but this teacup represents the culmination of one man's efforts to make the very first hard-paste porcelain in England: William Cookworthy. It was made around the time that Cookworthy was awarded a patent for producing hard-paste porcelain in 1768. England was nevertheless comparatively late to the porcelain party.

"The Germans were the first in Europe to independently discover how to make hard-paste porcelain."

The Germans were the first in Europe to independently discover how to make hard-paste porcelain, establishing the Meissen Porcelain Manufactory near Dresden in 1710. And over the next decade or so, detailed accounts of the

Small porcelain teacup with painted flowers and gilt decoration made by Cookworthy and Company, c. 1768–1770.

Chinese method began to emerge in Europe thanks to the writings of a French Jesuit missionary, François Xavier d'Entrecolles, who had learned the tricks of the trade in the city of Jingdezhen, China's porcelain-making centre. True porcelain, it turned out, required two key ingredients: a soft white clay called "kaolin" for the body, and a very fine stone powder called "petuntse" for the glaze – the "bones" and the "flesh" respectively, as Cookworthy referred to them. And it was Cookworthy who realised that equivalents of these raw materials could be found in England.

Born in 1705 to a poor Quaker family in Devon, Cookworthy worked as a chemist and druggist in Plymouth, but devoted himself to religious work, walking and experimenting with clay after the death of his wife, Sarah. His memoirs tell us that he discovered that kaolin and petuntse could be found "in immense quantities" in the county of Cornwall, in the far southwest of England. The "white, talcy earth" of Cornish kaolin was in fact already being used to mend furnaces for the region's most famous industry – tin-mining – but Cookworthy realised that it could equally be a valuable ingredient for porcelain manufacture. He first struck "white gold" in the mid-late 1740s on Tregonning Hill between Penzance and Helston, which had deposits of both kaolin and petuntse, before finding more further northeast at St Austell.

It would be about 20 years before Cookworthy was able to establish England's first hard-paste porcelain factory at Plymouth in 1768, during which time he experimented with perfecting his recipes and techniques. But persistence seems to have been a mark of his personality: as a teenager, he is said to have walked all the way from Devon to take up an apprenticeship at a London apothecary, a distance of some 300 km (about 200 miles). Cookworthy described in vivid detail the chemical processes he developed for preparing his hard-paste porcelain. To purify petuntse, for example, he recommended filling a crucible with a type of granite known locally as "Moor-stone", which was then "exposed to the most violent fire of a good wind furnace", causing the granite to melt "into a beautiful mass … one part of it will be almost of a limpid transparency [petuntse], and the other appear in spots as white as snow".

The fruits of Cookworthy's labours helped earn him a place in intellectual society. In 1768, Captain James Cook and naturalists Joseph Banks and Daniel Solander dined with him before sailing from Plymouth aboard HMS *Endeavour*, on the first of Cook's three famous voyages. It's tempting to think that they ate from tableware made from Cookworthy's very own hard-paste porcelain. His business moved to Bristol just two years later, where production continued for about a decade. The history of Cookworthy's hard-paste porcelain manufacture may be relatively brief, but the extraction of the raw materials continued well into the 19th century for uses in other everyday products like paper and paint.

Today, this industry has literally left its mark on the landscape around St Austell in the form of the "Cornish Alps": pointed peaks of waste minerals left over from the mining of china clay. So, next time you're enjoying a nice cuppa, remember that the piece of "white gold" you hold in your hands is part of a story thousands of years in the making.

Pink Apple iPod Mini, first generation, 2004.

IPOD

"In the iPod's 20-year lifespan, music technology and the way audiences choose to consume music has changed beyond all expectations."

The iPod revolutionised the way we consume music, popularising digital audio formats in its iconic, compact and colourful designs. The iPod Mini was introduced in 2004 as a more slimline and compact alternative to the larger, standard iPod, which was first sold in late 2001. The iPod Mini had a 4GB capacity that could hold around 1,000 songs, a (rechargable) battery life of around 8 hours and could be carried with you everywhere. Today, many of us soundtrack our lives via our iPhones or Androids with streaming services and digital playlists, rendering the iPod obsolete despite being the cutting edge of technology only a couple of decades ago. Arguably a victim of its own success, the iPod was discontinued by Apple in 2022, stating in press releases that "the spirit of the iPod lives on" in the way music is integrated across all their products. In the iPod's 20-year lifespan, music technology and the way audiences choose to consume music has changed beyond all expectations. It was the perfect transitional format between digital streaming and physical audio formats such as CDs.

For most of history, music was heard solely through live performances; someone singing, playing instruments, humming or whistling right in front of you. Or, if you were lucky, you could hear it via a mechanical musical instrument, like a music box or automata. Your favourite songs were preserved by oral tradition or with sheet music. That is, until the end of the 19th century, when it became possible to record and listen back to sounds for the first time with the introduction of the phonograph.

Invented in 1877, the phonograph played back sound that was recorded onto a piece of tin foil wrapped around a cylinder. Somebody spoke (or, more likely, shouted) into a mouthpiece, which was attached to a diaphragm, and the sound waves then caused the diaphragm to vibrate. The diaphragm was connected to a stylus that pressed into a cylinder covered in a thin layer of tin foil. Turning a handle made the cylinder rotate, and the stylus then cut a groove into the tin foil, leaving a recording of the sound. When the cylinder was rewound to the beginning, the stylus could then follow those marks, replaying the sound through a horn placed into the mouthpiece. The tin foil was eventually replaced with much sturdier wax cylinders,

The first phonograph played back sound that was recorded on a piece of tin foil wrapped around a cylinder – an innovative but time and labour intensive process.

thanks to the work of Alexander Graham Bell, inventor of the telephone, who created the "graphophone" – similar to the phonograph, but with wax cylinders – in 1887.

Although the quality of the sound would be poor to modern ears, it was groundbreaking at the time; those able to afford it could, for the first time, buy a device to listen to recorded sound and even record things for themselves in their own home. Edison's "all-wax" cylinder became a standard format, interchangeable between the phonograph and graphophone, and reusable because it was possible to shave down the wax and create a new recording. This made pre-recorded music more accessible, and marked the emergence of the music industry, with cylinders enduring as a dominant music recording format until the gramophone outpaced its popularity around 1912. Mass production of phonograph cylinders ceased in 1929.

Patented by German-American inventor Emile Berliner in 1887, the gramophone with its iconic brass horn took several years to become commercially available and was initially marketed as a toy. The gramophone was like the phonograph in many aspects, but Berliner made several improvements to quality and efficiency (and some necessary changes to avoid infringing Edison's phonograph patent). The most notable difference was its use of round discs instead of wax cylinders.

These discs were made of a much tougher and longer-lasting material than phonograph cylinders. The sound was recorded onto them with lateral incisions created by the stylus vibrating from side to side, instead of the phonograph's up and down motion. This created uniform, deep lateral grooves in the disc, which prevented jumping when the recording was played back (a problem with phonograph cylinders). These grooves also reproduced sound with a greater dynamic range. Berliner designed these discs with mass production in mind, unlike the phonograph cylinders that needed to be individually recorded. Copies of Berliner's disc records could be made by electroplating an original disc to create a "negative" with ridges instead of grooves, which could then exist as a stamp to produce copies en masse.

These discs remain very similar to their original design despite innovations over the years to increase quality, lifespan and structural integrity. They are still recognisable and used today in the form of LP (long-playing) microgroove records, which became the industry standard for gramophone discs when they were introduced in 1948 by Columbia Records. The introduction of the transistor radio in 1954 provided the gramophone with its most significant competition; the transistor radio was a pocket sized device, while the gramophone was a bulky piece of furniture. (Although their technical functions improved, record players remained big to accommodate the large size of LP vinyl.)

The history of radio goes back much further than 1954, however. The first transmission of sound over radio waves took place in December 1906, when Reginald Fessenden broadcast an hour of voice and music from Brant Rock, Massachusetts. Radios became available to the consumer market in the 1920s, and were a common sight within most homes by the 1930s. Initially, it was thought that by providing free access to music, radio would kill the phonograph and reduce the profitability of the music industry. The industry did see a reduction in profits, leading to copyright lawsuits across the world, which initiated the royalty payment scheme for the public playing of music. The American Society of Composers, Authors and Publishers (ASCAP) began collecting royalties in the US in 1923, and Phonographic Performance Ltd was formed by record companies EMI and Decca in the UK in 1934 to handle licensing.

While the transistor radio certainly changed consumer habits due to its portability (it is

The phonograph cylinder, created by Thomas Edison in 1877. His design would be improved upon by Alexander Graham Bell's laboratory in the 1880s.

CBS Laboratories chief engineer Dr Peter Carl Goldmark demonstrates a long-playing (LP) microgroove record produced for Columbia Records, 1948.

still used in many car radios today), the standardised LP disc provided consumers with the ability to choose what they wanted to listen to, and so maintained its popularity for decades. Some describe the 1970s and '80s as the "golden age" of the LP record. It was during this golden age that technological advancements started to ramp up, however, with shinier, newer formats rapidly introduced to replace the old. The music industry adapted to changing technology and tastes, with MTV launching a round-the-clock music video TV channel in 1981, bringing a new visual element to the industry. Additionally, advancements in record players, such as mixing decks and turntables, signalled the dawn of the "disc jockey" (or DJ), who plays recorded music in front of live audiences and on radio. Both MTV and DJs changed the culture of musical consumption, influencing the popularity of artists, songs and even entire genres.

Cassette tapes were invented in 1963, but they did not become a serious competitor to LP vinyl records until the mid-1980s. Tapes came in multiple formats, with eight-track tapes (recorded onto magnetic tape) being the main alternative to LPs for most of the 1960s and '70s. But they were no match for what was to come. The first "Walkman" – a portable audio playing device – was introduced with a cassette tape format in 1979, helping to make cassettes the dominant tape format. However, that same year the compact disc (or the CD) was invented, a format that did not have a tape that needed to be rewound manually (the "skip song" function wouldn't come until much later!). Once it became commercially available worldwide in 1983, the CD quickly usurped the cassette and the LP, becoming the most popular audio format by 1990 and remaining so until its peak in 2002.

While the CD emerged from the 20th century as the dominant mode of music listening, its popularity would begin to wane with the introduction of the iPod in October 2001. The iPod was not without competitors, though. The MP3 audio format was patented as early as 1989, but it didn't take off until internet usage became more widespread and digital audio players came to the market in 1997. The iPod was the first MP3 player to hold 1,000 songs and feature a 10-hour battery life, so other MP3 players on the market were no match for the iPod and iTunes. iTunes' user-friendly interface, the iPod's simple, colourful designs and Apple's marketing strategies helped the iPod dominate the digital audio format for much of the early 2000s.

The transition to digital audio formats saw an increase in unauthorised peer-to-peer sharing of audio files (also known as piracy), which led to numerous lawsuits over copyright infringement from record labels and artists alike – most famously with Napster, which was founded in 1999. This continued well into the 2000s, with apps such as LimeWire and websites such as Pirate Bay, where anyone could download any type of file (or a virus that would ruin your computer). LimeWire was shut down in 2010 after a legal battle against the Recording Industry Association of America, and the three founders of Pirate Bay served prison time in their home country of Sweden in 2009 for copyright infringement. While the website is still active, internet service providers have been ordered to block it in some countries.

Today, streaming reigns supreme, with Spotify claiming to be "the world's most popular audio streaming subscription service with more than 551 million users". In the US in 2015, streaming became the largest revenue source for digital music, overtaking downloads, CD and vinyl purchases. Worldwide charting was adapted to include streaming with the creation of an "album-equivalent unit" metric in the USA in the mid-2010s, in which 1 album sale equals 10 downloads or 1,500 streams of a song. While streaming is a cheaper, space-saving option for consumers, the format has been criticised for the way it pays artists, with many legal disputes hitting the news in recent years. British politicians have even called for a complete reset of streaming revenue systems, with the DCMS (Government Department for Digital, Culture, Media and Sport) Committee producing a report calling for reform in 2021.

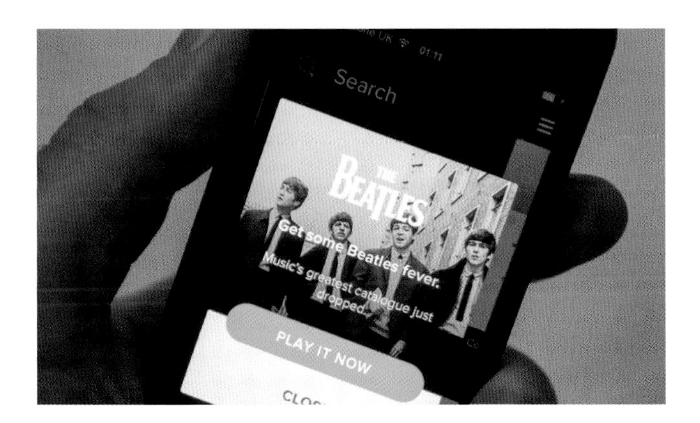

Spotify is the world's most popular music streaming platform, with over 550 million active users.

While modern audiences favour streaming, there has recently been a resurgence in "obsolete" musical formats. Vinyl record sales overtook CD sales for the first time since 1987 in 2023, and in 2021 CD sales increased for the first time in almost 20 years in the US. Perhaps people still like to own a physical version of their favourite songs, with streaming more akin to digitally borrowing the song to listen to. Or perhaps this is the consequence of nostalgia, or perceived differences in quality between analogue and digital media. Regardless, modern music consumers have a wealth of formats to choose from, and while the iPod isn't as ubiquitous as it was in the early 2000s, it is still a viable option for consumers – so long as the technology remains available to support it. It might no longer be possible to purchase one brand new, but with 450 million sold as of May 2022, the iPod might see a resurgence on the second-hand market, or perhaps even as the next big vintage collectible one day in the future.

STICKY TAPE

Glossy, matte or double-sided – sticky tape is a domestic staple. Utilised nowadays for a multitude of purposes, the origins of adhesive tape date back to 1845, when American surgeon Dr Horace Day applied a rubber adhesive to strips of fabric, essentially inventing the first surgical tape. The applications of sticky tape would go on to be instrumental in supporting the Allied efforts of World War II and winning two scientists the Nobel Prize in Physics.

The pressure-sensitive adhesive we know today was invented in 1925 as a solution for car manufacturers painting two-tone vehicles, a style that was trendy at the time. The painters found it difficult to achieve a sharp divide between the two colours, and the techniques used for dividing tones at the time usually resulted in a messy finish.

American engineer Richard Drew pioneered a solution to the painters' conundrum after overhearing their complaints while on a work visit to the auto manufacturers. At the time, Drew was selling sandpaper for a manufacturing company called 3M. He used a combination of crepe paper, cabinet makers' glue and glycerine to create the earliest version of what we know today as masking tape – a damage-free solution for paintwork.

Earlier versions of Drew's tape were complained about by the painters to be "scotch", a rather derogatory term meaning "cheap", as the prototypes didn't have enough adhesive applied to them. Ironically, the tape was branded as "Scotch" and the packaging embellished with a plaid pattern.

Drew went on to develop "cellophane tape" in 1930. Utilising cellophane (only recently developed by a Swiss chemist in 1908) as the backing for pressure-sensitive tape, Drew invented the world's first transparent tape. Cellophane tape was popular among bakers and grocers, presenting a neat and moisture-proof way to seal food in packaging that was also made from cellophane. Sticky tape rose in popularity in subsequent years, particularly during the Great Depression when money was tight and replacing broken items altogether was too expensive, it played a vital role in basic household repairs.

"Graphene is the world's first two-dimensional material . . . stronger than diamond yet a million times thinner than a single strand of hair."

In 2004, Andre Geim and Konstantin Novoselov found a reliable and cheap method for obtaining monolayer graphene flakes from graphite, using this tape dispenser.

In 1943, Vesta Stoudt, a mother from Illinois working at an American plant, recognised issues with the seals being used for ammunition boxes being used for the war effort. Soldiers would waste precious time scrambling to open the boxes and Stoudt, having two sons in the US Navy herself, came up with a solution for a new kind of tape that could be torn by hand, rather than cut with scissors. She wrote a letter addressed to the president himself, suggesting a new kind of "waterproof cloth tape" she had tested that could be used to seal the ammunition boxes. Former president Roosevelt approved Stoudt's invention and the production of duct-tape was undertaken by a company called Johnson & Johnson. Tape was vastly utilised by the war efforts of the 1940s, aiding in sealing, insulating and holding together materials. The industrial and household applications of sticky tape exemplified its versatility as an adhesive, but this was only the beginning of what sticky tape was to achieve.

Sample of concrete enhanced with 0.1 per cent graphene after compression strength testing.

In 2004, during an evening of unconventional experimentation at the University of Manchester, two scientists isolated the world's strongest material using none other than the very roll of Scotch sticky tape on page 193. "Friday Night Experiments", hosted by Dutch-British physicist Andre Geim, were experimental laboratory sessions held to encourage out of the box thinking about old problems.

Geim and fellow physicist Russian-British Konstantin (Kostya) Novoselov's team were working with graphite in an attempt to solve an age-old enigma. Graphite, a naturally occurring form of carbon found in metamorphic rocks, is made up of sheets of carbon atoms where each individual sheet measures just one atom in thickness. These single atomic plates of graphite are known as "graphene", the first two-dimensional material to be discovered. Since identifying its existence, scientists had been attempting to isolate graphene for decades, employing all sorts of techniques to reach the atom-thick structure. However, to no avail. It was thought that graphene was simply too unstable and minuscule to isolate. Geim and Novoselov pursued this challenge with fresh thinking.

When preparing graphite for viewing under a microscope, it is common practice to clean the samples with sticky tape, effectively removing any residue from the material's surface that may obstruct the experiment. A member of the team was curious about the sticky tape being used to clean the graphite and questioned whether anyone had observed the flakes of graphite stuck to the tape itself. After retrieving a piece of tape from the bin and inspecting it under a microscope, they noticed that the tape did indeed have some thin flakes of graphite stuck to it. Some of them were so thin, in fact, that they were even transparent. By continually separating these flakes, Geim and Novoselov were able to obtain fragments that were just one atom thick, successfully isolating graphene for the first time.

Graphene is the strongest, thinnest and most conductive material in the world, leading to infinite new possibilities in our everyday lives. Geim and Novoselov's technique came to be known as the "Scotch Tape Method" and, along with subsequent experiments demonstrating the electrical properties of graphene, would win Geim and Novoselov the Nobel Prize in Physics in 2010.

Carrying on the tradition of reviving old hypotheses, in 2008, American researchers at the University of California, Los Angeles (UCLA), confirmed that sticky tape could even produce x-rays. Theorised as early as 1953, scientists in Russia suggested that sticky tape, when peeled in vacuum conditions, released a burst of energy in the form of x-rays. The team at UCLA was able to show that the x-rays produced were strong enough to leave an x-ray scan of a finger on photographic film. This phenomenon is known as triboluminescence, when light is generated from friction, and so sticky tape continues to act as a tool for making great advances in science.

Sticky tape may determine how affordable and accessible x-ray scans are in the future, or how sustainably we can develop existing technology using the "wonder material" graphene. Drew's innovative thinking eventually led his boss at 3M to develop a "15 per cent culture" of innovation, where employees at the manufacturing firm are encouraged to use 15 per cent of their working hours to think creatively. When an ordinary tool is utilised by extraordinary thinking, great feats can be achieved, and thinking outside the box and challenging the status quo can lead to remarkable discoveries ... and it could all start with a few playful experiments fuelled by curiosity and some sticky tape!

Coating the filament in graphene makes a lightbulb more efficient – and also cheaper to manufacture.

SWISS ARMY KNIFE

The Swiss Army Knife is an accessible tool for anyone attempting to take on the wilderness, even for those who shudder at the thought of spending a night in a tent. Containing an array of tools from a knife to nail clippers, a wood saw to wire cutters, a corkscrew to a chisel, it is the ultimate multifunctional portable tool.

Its unique design has been replicated by other manufacturers trying to ride off the back of its success, though there are only two official manufacturers of the Swiss Army Knife: Victorinox, the first, acquired the second, Wenger, in 2005. Victorinox was founded by Karl Elsener I in 1884, when he opened a knife cutlers' workshop in the village of Ibach, Switzerland. After establishing the Association of Swiss Master Cutlers in 1891, Elsener began supplying the Swiss Army with knives, and patented his original Schweizer Offiziers- und Sportmesser (Swiss Army Knives and Sporting Knives) in 1897. He later named the company after his late mother, Victoria, registering the brand name along with the cross and shield emblem.

The company's name also nods to a major innovation that occurred in the early 1900s: the creation of stainless steel. The development of the material began in the 1800s when scientists experimented with iron alloys (mixtures of metals containing iron). These included an iron-chromium alloy, which was found to be more resistant to weathering and degradation. When iron atoms in an iron-containing metal react with oxygen, they form iron oxide, also known as rust. The addition of chromium prevents this from happening. Instead of reacting with the iron, the oxygen reacts with the chromium, creating a very thin protective layer. Stainless steel can still rust, but it is far more resistant, prolonging its useful life. The introduction of stainless steel to industry, including to Elsener's factories, was so impactful that the company integrated the French word for the material, *inox*, into its name Victorinox.

"The name Victorinox nods to a major innovation that occurred in the early 1900s: the creation of stainless steel."

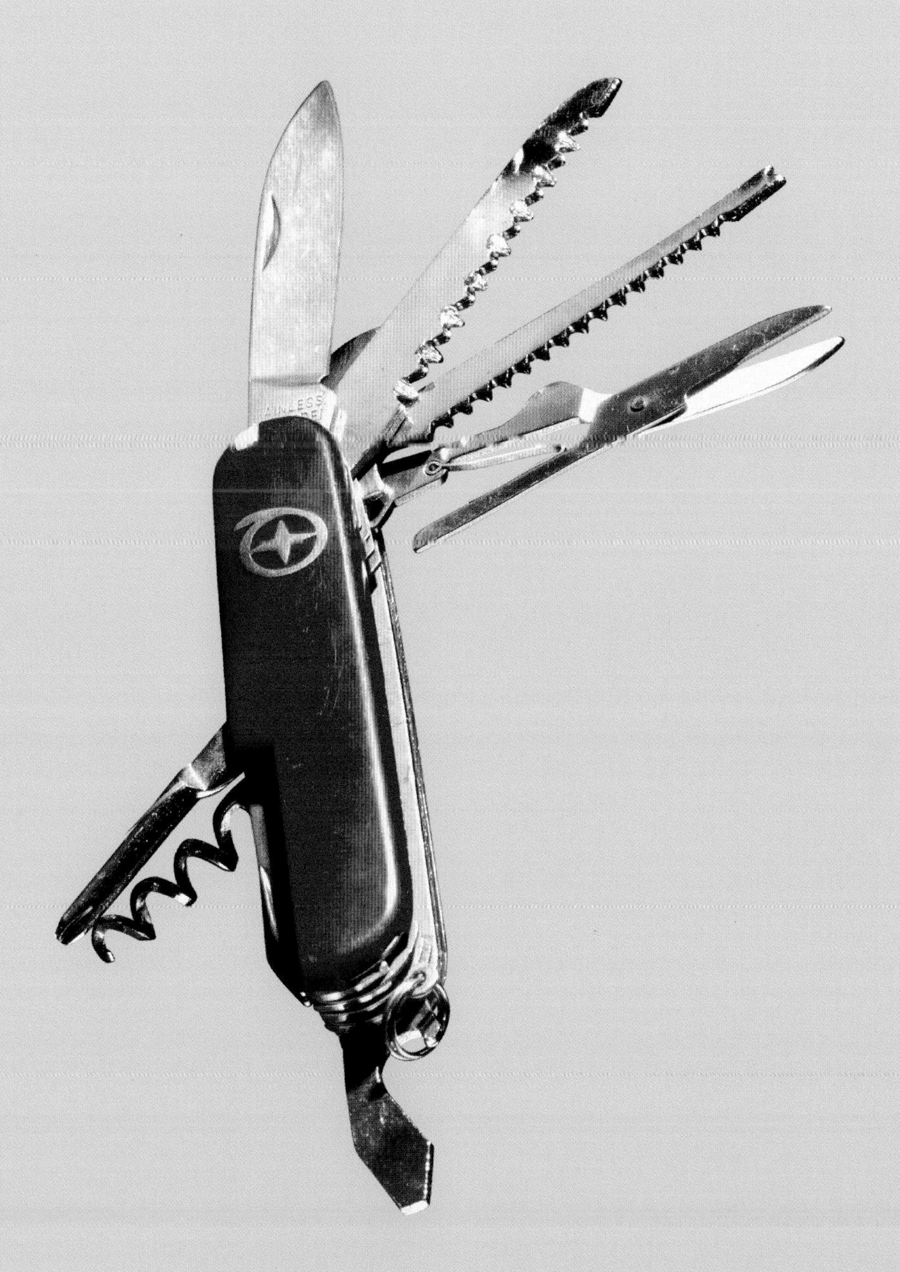

Pocket knife with eleven blades made around 1998, consisting of tools including a screwdriver and a plastic toothpick.

The Swiss Army Knife is not the oldest multifunctional portable tool. The innovative Romans had their own version, with one of the best examples kept at the Fitzwilliam Museum in Cambridge. Dubbed the "Roman 'Swiss Army Knife'" by the museum, the utensil is a folding eating implement with a three-pronged fork, a spatula, pick, spike and knife. While we can't be certain of how it was used, it is thought that the spatula might have been employed to get sauce out of narrow necked bottles, like garum, the infamous fermented fish sauce. The pick may have come in handy as a toothpick, and the spike could have proved useful at getting the meat out of snails. Many similar Roman tools have been found, but they are less elaborate and made from bronze, unlike the one held by the Fitzwilliam Museum, which is made from silver. The different materials suggest that these tools were accessible to different groups of people, with luxury versions available to those with money to spare.

But even the Romans were not the first to have invented clever portable tools. While we may think of the Stone Age as a homogeneous period, it was made up of many different eras with complex cultures. Our ancient ancestors were anatomically modern humans with all the cognitive creativity we have now, and even our long-gone cousins, the Neanderthals, are now believed to have had intricate social systems with well-crafted tools.

"The Swiss Army Knife is not the oldest multifunctional portable tool. The innovative Romans had their own version, a folding eating implement with a three-pronged fork, a spatula, pick, spike and knife."

The Mesolithic, or Middle Stone Age (approximately 20,000 to 10,000 years before the present in the Middle East, and 15,000 to 5,000 years before the present in Europe), saw major innovation in tools. Microliths – minuscule, chipped stones usually no more than 1 cm in length and 0.5 cm in width – became common during this era. Despite their small size, they were effective at carrying out a range of tasks. Ancient humans crafted these tiny tools to process plant material and meat, to work wood, and probably to make clothing. Archaeologists have discovered a wealth of evidence showing that microliths were not used in isolation but were part of larger objects. Glued into bone or wooden notches using plant resin, these microliths could be assembled to make composite tools. Not only were these tools versatile, but they could also be mended when broken by replacing the damaged stone with a new one. Archaeology shows that humans have always had a desire for efficiency and versatility, and a creative spirit that has proved central to our survival.

Today, the Swiss Army Knife has been able to maintain many of its original design elements from the late 1800s, such as can openers and screwdrivers. You can also opt for an old-school wooden casing. But how is this tool adapting to the technology of the 21st century? Victorinox has already created pocket knives that contain a 32GB USB drive, though further digital elements have yet to be

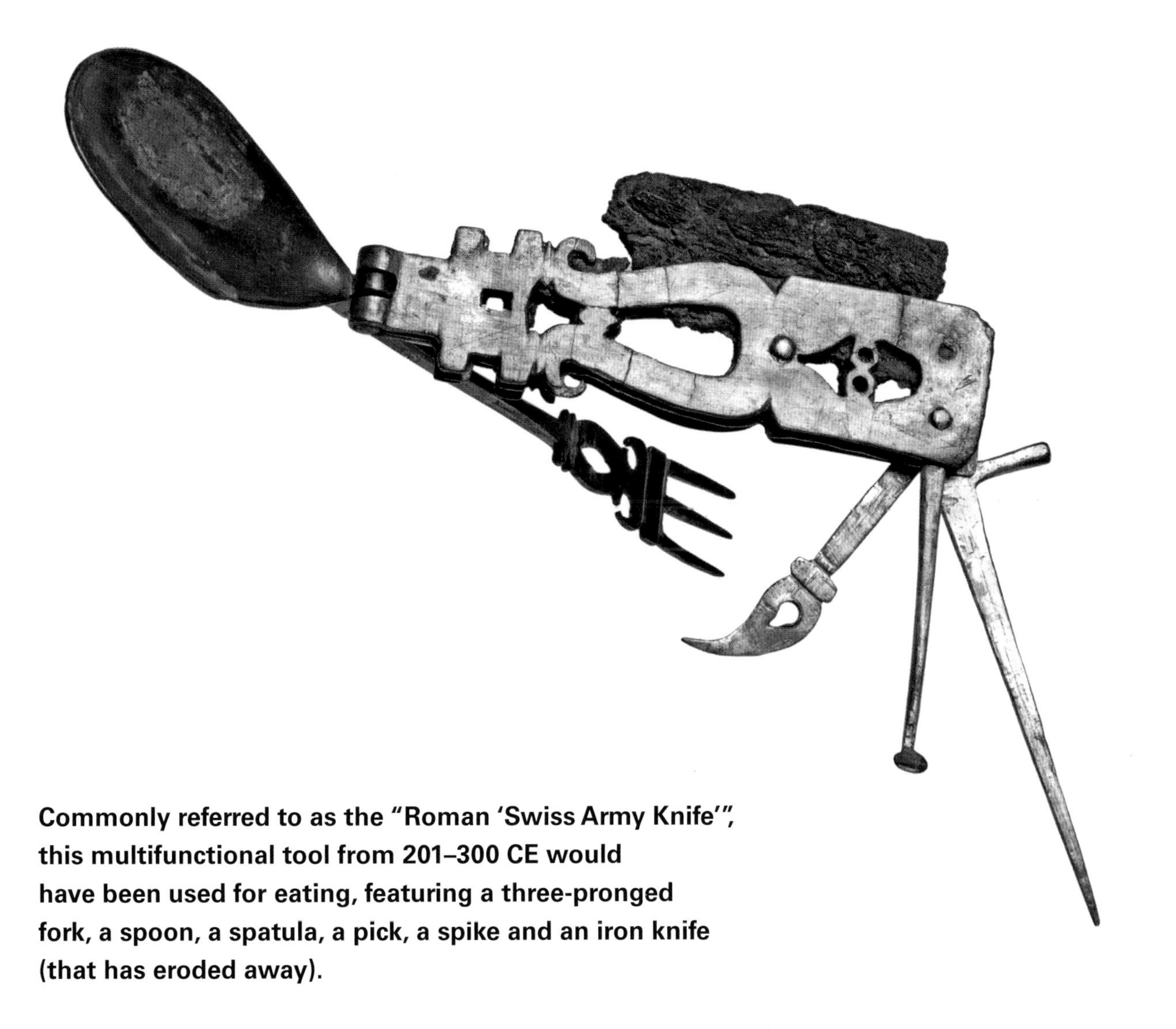

Commonly referred to as the "Roman 'Swiss Army Knife'", this multifunctional tool from 201–300 CE would have been used for eating, featuring a three-pronged fork, a spoon, a spatula, a pick, a spike and an iron knife (that has eroded away).

added. Its analogue focus does not mean that the company is stagnant, however. As well as designing specialised tools for hobbies and tasks, such as the "Victorinox Picknicker" for outdoor dining, and the "Garden" series of knives, the cutlers have branched out into other products, offering watches, travel accessories and even fragrances.

To what extent does the Swiss Army Knife really need to embrace our increasingly digital lives? Perhaps it is beloved precisely because of its off-line functions – for its use in activities that take us away from screens, be it gardening, camping or picnicking with friends. While a pruning knife that can simultaneously cut delicate shoots and offer digital feedback on your plant's nutrient levels might have its uses in future, a good quality pocketknife will do the trick for many. And when you're hiking in the wild without charging points or cellular signal, a trusty corkscrew and a can opener might prove more essential than a smart watch.

QWERTY KEYBOARD

If you live in an English-speaking country, chances are that you are writing your emails or WhatsApp messages on a QWERTY keyboard (or a variation of it, if you are typing from France, Italy or Germany). But have you ever wondered how it is that most of the world ended up using the same keyboard arrangement? Named after its first six letters, the history of the QWERTY keyboard is an exciting story of technological development, failure, adaptation and, most importantly, the social and material dynamics of innovation.

It is said that the typewriter was invented 52 times before a feasible prototype was possible. There is evidence that dates typing instruments to the 16th century, however, the typewriter that paved the way for the ones that we know today was developed between 1867 and 1873 by Milwaukee-based printer and tinkerer, Christopher Latham Sholes. The amount of clerical work that the new industries were producing would have been known to retired printer Sholes, and it is believed that he decided to venture into creating his own after reading an account of a "type writing machine" in *Scientific American* in 1867. He was assisted by Carlos Glidden and Samuel W. Soule (who was later replaced by the investor James Densmore). The prototype of the "Type Writer" machine, as they called it, had a 28-key, piano-like keyboard arranged alphabetically, which resembled the Hughes-Phelps printing telegraph. However, it wasn't until 1868 that Sholes and Glidden filed the patent for their perfected model of "Type Writer". The patent model still had a piano-like keyboard encased in a wooden frame, which made the machine heavy and difficult to operate. In 1870, assisted by mechanical engineer Matthias Schwalback, Sholes prototyped a keyboard with 38 keys arranged in 4 rows. In 1872 he introduced a 42-key one. While Sholes was working out the different keyboard layouts of the writing machine, he was also discussing the sale of the manufacturing rights to firearms maker Remington & Sons – a deal that was agreed in 1873.

"It is said that the typewriter was invented 52 times before a feasible prototype was possible."

Sholes and Glidden patent typewriter original model, 1875.

There are multiple theories about how the QWERTY keyboard came to be the universal one. In 1873, the manufacturing of the typewriter went to Remington & Sons' sewing machine division, which explains why the first Remington typewriters included features similar to the sewing machine, such as a stand with a treadle. One theory argues that in the hands of Remington's engineer, Jefferson Clough, and manager, William K. Jenne, Sholes' earlier four-row button keyboard design that followed an almost alphabetical order and numerals was changed in order to slow down the operator, thereby avoiding the jamming and colliding of the typebar when consecutive letters were pressed. The jamming problem of the keyboard encouraged Remington to investigate the most used letter combinations in the English language, which led them to place frequently occurring letters on opposite sides of the keyboard. It was at Remington that the first variations of the QWERTY keyboard were created. Remington replaced the full stop of the early QWE.TY version with the letter R, and moved the Y to the right-hand side of the keyboard. However, it is said that Sholes demanded Y to be moved back to the centre of the keyboard, creating the QWERTY order.

An alternative theory suggests that one of the reasons behind the QWERTY keyboard was that "Type Writer" could be easily typed using a single row of the keyboard – the ultimate marketing strategy of the era. Other researchers have suggested that the decision to arrange the keys in the QWERTY arrangement was informed by Morse code testers, who needed to quickly transcribe the received messages. Regardless of the various theories on how the keyboard came to be and who was responsible for influencing the change (the machine mechanics or the machine users), it is a fact that in 1874 the Remington 1 was the first mass-produced typewriter to include the QWERTY keyboard.

Despite the typewriter keyboard becoming an indispensable part of modern technology, it was hardly a success when it first came out. The Remington 1 cost $125 (over $3,000 today), a hefty price for a technology whose wider need had yet to be established. With only some modest sales on its first model, Remington released the Remington 2 in 1878. The second version of the writing machine had 40 keys and included upper- and lower-case options. In 1882, the last modification to the QWERTY keyboard was made by switching the letters C and X to where they sit today. By the 1890s, it is estimated that around 100,000 QWERTY typewriting machines were being used in the US, and user training had been taken up by several companies, among them Remington itself. This all culminated in 1893, when the five major typewriting manufacturers agreed to make QWERTY the standard.

The preference for the QWERTY keyboard, however, was not due to its efficiency. While QWERTY was being dubbed the universal keyboard by manufacturers, other companies were still producing alternative keyboards. The DHIATENSOR or "Ideal" keyboard used by typewriting companies Hammond and Blickensderfer, for example, consisted of a three-row keyboard and was much smaller, lighter and easier to manufacture than Remington machines. Despite how efficient alternative machines and keyboards might have been, businesses were reluctant to invest in new keyboards due to a lack of operators, and in turn operators didn't want to retrain in new layouts. Because businesses and operators were resistant, manufacturers were reluctant to release new variations and could easily retrofit new layouts into existing machines.

At the time all this was happening, Remington was already offering typing courses for the QWERTY. These courses were in most cases aimed specifically at women, as the introduction of the typewriting machine brought with it the introduction of a new set of users. With the advances of the Industrial Revolution and capitalism, workplace bureaucracy had increased significantly, which meant that clerical work was in high demand. Remington's typing courses were introduced as early as 1875 with many other schools following the trend. Middle-class women were seen as the solution to inexpensive educated labour in the office and because typewriting was compared to sewing and playing piano, it was considered an acceptable and feminine way of making a living. Despite the persistent gendered division of labour, with women relegated to clerical work and men in decision-making positions, the typewriter afforded some economic independence to women and became a symbol of the "new women."

For decades, few manufacturers attempted to offer alternative keyboard layouts until 1932, the year American Psychology Professor August Dvorak and educator, W.L. Dealey proposed the "Dvorak Simplified Keyboard" (DSK). Dvorak argued that the new layout, a four-row keyboard using a similar combination to that of the "Ideal", required less finger motion and reduced typist strain because it enabled the most used English combinations to be written using the home row (where fingers naturally rest on the keyboard). During World War II, Dvorak became a lieutenant in the US Army, which gave him an opportunity to empirically test the efficiency of the DSK. After a short trial, he argued that the cost of retraining typists in the new layout could be quickly recovered by the increased efficiency of the keyboard. However, despite Dvorak's best attempts to convince businesses, manufacturers and typists, the new layout didn't take off.

Today, despite most devices having a QWERTY keyboard by default, most operating systems, such as macOS, Microsoft and Linux, offer the option to switch to the DSK layout and, albeit quietly, the DSK's claims of efficiency can still be heard. However, QWERTY remains the norm. Since its commercial release in 1874 its use has reached far beyond the typewriting machine and the computer, and expanded into the realm of smartphones and other digital devices that have little need for it as four-finger typing has been replaced by two-finger (thumbs) typing and autocorrect. In understanding how this happened, most research focuses on establishing whether it was the new users who influenced the layout of the keyboard, such as telegraphers or stenographers, or whether it was the mechanics of the typewriter, e.g. the clash of the bar while typing. However, like all technological developments, the answer does not lie in just the social or the material, but somewhere in between. QWERTY came at a time when typewriting machines were a new, unknown technology, and its uses and users were yet to be established. There was no standard as much as there were no experts, so the road was free to be paved by the relationship between humans and technology.

BATH TOYS

In the depths of the Science Museum is a treasure trove of objects not on public display. Wandering through the Oceanography section, you might marvel at the ingenious instruments and curious contraptions that scientists have used to observe, measure and understand the seas for centuries. And in one particular box, you'll be surprised to find a small family of plastic ducks, turtles, beavers and frogs staring back at you. Bath toys? What could these rudimentary objects have to do with the science of our planet?

In fact, these toys are unusually intrepid, earning their place in the national collection due to their adventures in Earth's biggest bath. They, and others like them, began to mysteriously wash up on the Alaskan coast in November 1992. By the following summer, around 400 had reached the shore. Research by intrigued journalists revealed that they had originated from a container aboard the cargo ship *Ever Laurel*. While traversing the North Pacific in January 1992, en route from Hong Kong to the US, the ship was caught in a storm and 12 of its containers were washed overboard. Damaged by the waves, the containers ejected into the sea their contents, including almost 29,000 Chinese-made "Friendly Floatee" bath toys.

When he heard that dozens of plastic creatures were making landfall, Seattle-based oceanographer Curtis Ebbesmeyer saw an opportunity. For decades he had experimented with techniques for making visible complex ocean currents, knowledge of which could help predict the movement of oil spills. He tried dropping purpose-built marker buoys into the sea, and tracking flotsam and ocean debris, including trainers and hockey gloves that had escaped from shipping containers. To Ebbesmeyer, the Friendly Floatees were not a curiosity: they were data. Along with his collaborator, fisheries scientist Jim Ingraham, Ebbesmeyer gathered information about when and where the toys reached the coast. The scientists plugged this into a computer model they had developed to simulate currents at the ocean's surface.

"The Friendly Floatees were not a curiosity: they were data."

Ebbesmeyer's fixation with drifting objects as a source of knowledge was nothing new. This technique is thought to date back many thousands of years, especially in

A "Friendly Floatee" duck, one of 29,000 lost overboard in a storm in 1992.

cultures reliant upon the ocean for food and transport. In the 19th century, Western scientists released tranches of miniature glass bottles into the sea, inviting anyone who found them to report their whereabouts using the instructions contained within. From the 1970s, scientists began using satellites to track specially designed drifting buoys from space. However, in comparison with higher-tech methods, an advantage of the Friendly Floatees was their sheer number, which yielded a rich suite of data with which to test and refine ocean models. Ebbesmeyer cultivated a network of enthusiastic beachcombers to scour the shoreline and notify him of their findings.

The Friendly Floatees have, for example, contributed to our understanding of gyres – vast, circular ocean current systems. The toys were swept into the North Pacific Gyre, which runs clockwise between Japan and the west coast of the US. Sightings on the shoreline were more common in some years than others, which helped oceanographers establish that the gyre completed a cycle about once every three years. The range of the toys' landing locations also confounded simplistic understandings of circulation patterns. Using their model, Ebbesmeyer and Ingraham predicted some toys would have escaped the gyre and drifted northwards through the Bering Strait to become trapped in the Arctic ice pack. Others are probably still trapped in the gyre, and may never reach the shore.

There is a darker side to the tale of the Friendly Floatees, however. In recent years, we have become increasingly conscious of the problem of plastic pollution in oceans, with the toys themselves testament to the very durability of plastic. The sun and seawater may have taken their toll on the red and yellow dyes that originally coloured the beavers and ducks (though, interestingly, not the blue turtles or green frogs), but otherwise the toys remain remarkably unscathed from their epic journey.

Every year, at least 14 million more tonnes of plastic enter the ocean than the previous year, and the material takes hundreds of thousands of years to break down. Ocean gyres cause plastic waste to aggregate into massive garbage islands. But as horrifying as plastic floating on the surface of the ocean might be, even more lurks invisibly underwater. As it degrades, plastic is ingested by marine animals, causing harm not only to marine ecosystems but also, in turn, to us. The chemicals added to plastic to give it useful properties, such as transparency or flexibility, are present only in minuscule quantities in the products we use, but accumulate in the bodies of animals that eat ocean plastic. If consumed in sufficiently high concentrations, they can have harmful effects on human health.

Knowledge of the ocean is essential in addressing a related environmental challenge: climate change. Ocean currents are at the heart of the global circulation models that enable scientists to explore the possible climate futures we might face in the decades to come, depending on the actions we take today. Insights from, for example,

"The Friendly Floatees helped oceanographers establish that the gyre completed a cycle about once every three years."

These examples of Friendly Floatees give a clear indication of the sun's impact on the yellow and red dyes used on the beaver and duck toys.

the vast international Argo project, which generates a detailed profile of the ocean using thousands of pre-programmed floats, are helping us understand the forces that drive and shape our climate. In turn, this can help us build resilience to future change. Increasingly, autonomous underwater vehicles – ocean-going robots – are at the forefront of oceanographic research too, shedding light on places that are difficult to reach with conventional instruments.

Yet, people strolling along shorelines around the world remain important sources of information alongside these scientific advancements. Cornwall resident Tracey Williams became fascinated with the journeys of ocean plastic after discovering Lego washed up on her local beach. She has since used her own findings, and her network of tens of thousands of fellow beachcombers around the world, to contribute to scientific studies of ocean currents.

And for humble bath toys with unfulfilled wanderlust, the horizons continue to expand. In 2008, NASA scientists deployed dozens of rubber ducks in an attempt to trace the path of a glacier in Greenland. Six years later, students from the University of York sent their "Astroduck" toy to the edge of space attached to a weather balloon. Who knows what other adventures could be in store beyond the bathtub?

"Kenwood Electric Chef" food mixer, model A700,1950–1956.

KENWOOD A700

Kitchen appliances change the way we cook, eat and rest. While new products and cutting-edge technologies regularly enter the market today, even the Victorians had an appetite for kitchen gadgets. These were largely restricted to the middle and upper classes, but Victorians with spare money could purchase products like weighing scales, meat mincers and kettles. However, it wasn't until the Second Industrial Revolution in Europe and the US at the turn of the 20th century, when electricity was widely accessible, that appliance ownership began to boom.

In 1950, what would become a kitchen icon burst onto the domestic scene. The Kenwood Chef A700, a stand mixer with a multitude of attachments, was introduced to the British public at the *Daily Mail* Ideal Home Exhibition by the Kenwood Manufacturing Company. Company co-founders Kenneth Wood and Roger Laurence had worked as engineers in the British Royal Navy during World War II. Together, they set their sights on giving kitchen appliances a style and engineering update. Their first appliance was the 1947 "Turnover Toaster", which toasted both sides of bread simultaneously. In 1948, the company produced the A200, the precursor to the A700, boasting four speeds, a juicer, a mincer and two beaters.

While Wood and Laurence did not invent the stand mixer – that was Herbert Johnson, an American engineer and creator of the KitchenAid, in the 1910s – their innovative design is what set their mixer apart. Combining the best bits of the A200, and borrowing style ideas from their competitors, they created the A700. The vertically mounted motor allowed for a liquidiser to be placed above, and a series of gears provided a slow-speed outlet at the horizontal end, perfect for blending and mincing.

"In 1950, what would become a kitchen icon burst onto the domestic scene."

The A700 was endorsed and demonstrated by popular home economist Marguerite Patten, presenter of the BBC's *The Kitchen Front* during World War II, at London department store Harrods in 1950. The "planetary action" of its beaters – when the beater and its socket rotate in the opposite direction – made sure not a smidge of batter was left behind in the mixing bowl. In post-wartime Britain, with food rationing still in place, no morsel of food was wasted. The A700 made such an impression on customers that Harrods had sold out of them within a week.

After World War II, more women had opportunities for paid employment outside of the home, though they earned a fraction of what men did and were still expected to shoulder the majority of domestic labour, including cooking. Having a (seemingly) perfect domestic life was a sign of success in certain social circles, but with so much work to do for one person, this was often an uphill struggle. As such, key to the Kenwood Chef's desirability was its promise of saving precious time.

Even though women were beginning to earn their own money during the post-war period, the high price of the Kenwood Chef (equivalent to more than £600 in 2023) meant that it was out of budget for many. Expensive items were available on hire purchase, whereby people could hire out items and eventually own them through paying in instalments. But women were not legally allowed to sign off on payments themselves, and had to get a male member of their family to sign as a guarantor. This was one frustrating form of financial sexism that persisted in the UK until 1975, when the Sex Discrimination Act was passed.

As a result, wives would often have to convince their husbands to help them buy items targeted at women. They did, however, make popular gifts for husbands to bestow on their wives, as was exemplified by advertising slogans such as "The Kenwood you bought me prepared this wonderful meal", and "Shsh... we're giving Mummy a Kenwood Chef" (the latter advert including a "note to husbands" with payment information). While there was no

Kenneth Wood presenting one of his Kenwood Chef food mixers to model Jill Holt at the *Daily Mail* Ideal Home Exhibition, London, 1957.

doubt that women were the intended users of the Kenwood – slender, manicured hands touted the efficiency and speed of the mixer – these adverts implied that men still held control over financing and would benefit from the purchase.

The actual efficiency of these supposed time-saving appliances has been debated. Ruth Schwartz Cowan, science historian and author of *More Work for Mother: The Ironies of Household Technology From The Open Hearth to the Microwave* (1984), argues that the amount of labour done by women didn't really reduce because the expectations of labour changed alongside technological developments. The nuclear family was the norm, with the husband at work and the children at school, leaving the mother to maintain the household alone. The uptake of these domestic gadgets contributed to the idea that housewives were now capable of doing even more work, and to a higher standard than before. In essence, the goalposts shifted.

As well as being efficient and durable, the Kenwood Chef A700 was eye-catching, and the newly affluent middle classes were keen to show it off on the kitchen counter. Having an immaculate kitchen with the latest gadgets was key to the aforementioned aspirational lifestyle. Despite rationing in the UK continuing after the war until 1954, four years after the Kenwood mixer came out, dinner parties were one way in which socially minded hosts could demonstrate their domestic prowess. The events were less about serving up mouth-watering meals (who wouldn't salivate at the sight of a lime jelly tuna salad?), and more about keeping up appearances.

The Kenwood A700 is no longer manufactured, though some might be lucky enough to have recent editions or second-hand stand mixers. Its successor, the A701A, designed by Sir Kenneth Grange, was released in 1962. It shed the curvy design for a playful, sleek look, which was maintained in the series throughout the 20th century. The real achievement of the A700's design legacy is seen through the continuation and evolution of the chef's accessories – still enabled by the vertical motor. More than 70 years since Kenwood released its first mixer, equipped to mix, mince and juice, current models offer the same attachments in addition to many more. A lasagne roller, spiraliser and frozen dessert maker show the changes in our dietary desires.

Unlike in previous generations, having your friends over for a meal nowadays is less about showcasing your class aspirations and more about bonding. Cash-strapped and time-poor, younger generations are more likely to order takeaways together, or each bring dishes to share. For many with small kitchens, bulky baking stand mixers may be less favourable than smaller everyday appliances.

Even if you're not willing to fork out for a fancy stand mixer, household appliance crazes continue, whether it's blenders selling you the idea of quick-fix nutritious smoothies, or multifunctional air fryers. Icons like the Kenwood were innovators, but also succeeded because they knew how to sell lifestyle dreams in the form of consumer durables. Our economies are ever-changing and with that, so are our ideas of success and how we present them.

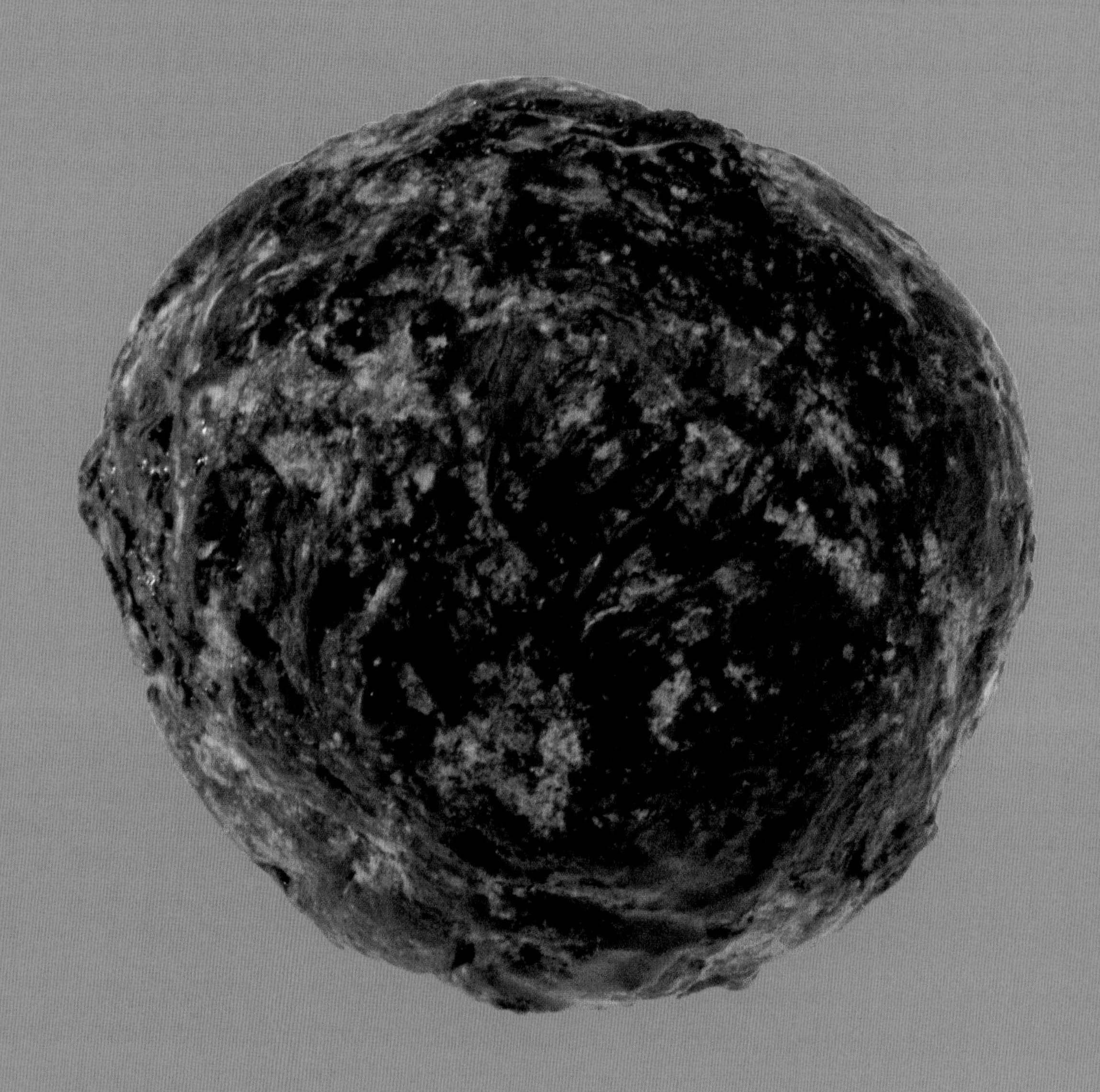

This rubber ball from Peru is thought to date to the 16th century.

PERUVIAN RUBBER BALL

The deceptiveness of the Peruvian rubber ball opposite is part of its initial appeal. Although it belonged to a child, it doesn't look like a children's toy. At first glance, it might be mistaken for a pebble, a rock or even a model of a murky planet.

But its initial deceptiveness also proves to be revealing. While this particular ball is thought to date to the 16th century, such balls have been central to the ancient cultures of what are now Central and South America for thousands of years. As representations of the sun and moon, they occupied a central place in the cosmology of Mesoamerican societies.

The anthropologist Robert W. Henderson has argued that the ball has been used since prehistoric times across all societies as both a fertility symbol and a way of staging mock combat between good and evil. The Mayan demigods, Hunahpu and Xbalanque, supposedly rescued the corpses of their forefathers by defeating the gods of the underworld in a game that was something like a mixture of volleyball and football.

More broadly, it appears these rubber balls were involved in games that saw two teams kick, volley, head or strike the ball across a central dividing line at high speed. The ball was heavy and would have been both dangerous and difficult to control with bare feet. The current popularity in South America of games such as futsal, which borrows conventions from basketball, handball and water polo as well as football, attests to the deep resilience of this cultural legacy.

"The behaviour of rubber balls represented the loosening of hierarchies and the reversibility of fortunes."

Over the course of some 3,000 years, the social meanings attached to these rubber ball games changed across societies and places. Even so, the game appears to have had widespread social value. It was played by everyone. The behaviour of rubber balls represented the

A ring for ball games at the outer wall of the ruined Mayan city of Uxmal in the Yucatan, Mexico.

loosening of hierarchies and the reversibility of fortunes – up and down, earth and air, life and death.

The rubber ball game was also tied to complex systems of time. The Aztec version of the game was scheduled according to the observable cycles of the planet Venus, while elsewhere the game was a focus for betting, a substitute for war or a prelude to ritual slaughter.

In Peru, there is evidence that infants were frequently killed in the hope that such offerings would ward off the cyclical El Niño of wet and stormy weather that periodically ravishes the north-western coasts of South America. The ball on page 212 was found in the grave of a child. It is possible that, while it may have been a symbolic gesture towards the afterlife, this ball belonged to a child killed as part of a similar sacrifice.

In the case of South and Central American societies, rubber balls like this one were not only part of intricate belief systems, but were also supported by a particular material culture. We think of vulcanised rubber as an invention of the Victorian era, but centuries before in Central America, latex from rubber trees was mixed with sulphur from the roots of the morning glory plant to achieve an organically hardened equivalent.

The resilience and regularity of rubber balls allowed for the emergence of games of great skill. For example, in contrast with South

America, early "football" in the UK was played between villages with an inflated pigs bladder. Not only did these "balls" bounce unevenly but they tended to collapse, often bringing games to a contentious and arbitrary end.

Indeed, it is no coincidence that the conventions of modern football were developed in England in the decade after Charles Goodyear patented vulcanised rubber in 1836. The development of the first rubber footballs in the mid-19th century not only made the bounce of the ball more predictable, it also enabled the creation of a set of standards for shape and size. Although the rubber cores of FIFA-approved balls are now laboratory tested to maximise the transfer of energy from the kicker to the ball, the measurements that define an official football have barely altered since 1872.

However, long before the practices and equipment of modern sport in Europe were transformed by the Industrial Revolution, fast-paced rubber ball games were being played in temples or on long, thin courts throughout Central and South America. For thousands of years, players built their strength and improved their reaction times by training with bouncing rubber balls like this one.

In 1521, the arrival of the Spanish conquistadors led to the violent suppression of the original rubber ball games. To the conquistadors, the elasticated rubber ball appeared to possess non-human powers. It oscillated between opposite conditions. It seemed to absorb, store and release energy. They considered it demonic.

Moreover, the Spanish understood that the ancient rubber ball games provided the physical and symbolic fulcrum of an entire culture. Priests and spectators watched the frenetic and dangerous games from above the courts, which were surrounded by high walls decorated with bright murals, where the unruly motions of the ball seemed to delay, relay and release time.

While the civilisational importance attached to these rubber ball games might seem difficult to empathise with, there is a branch of contemporary philosophy that is more sympathetic to the epistemological importance that these societies placed on the ball.

Modern thinkers have taken the ball to be a significant emblem of the "quasi-object" – that is to say, an object that is not quite an object because it oscillates between objectivity and subjectivity. The ball is an *it*, but also an *I* when sitting at the foot of a specific player, and a *we* when woven by a team into a series of passes. The ball makes it difficult if not impossible to separate the *it*, the *I* and the *we*, because these categories are always emerging out of each other. While we might now associate such thinking with advances in reality-warping scientific fields such as physics, we perhaps should not be surprised that the ball has often been understood as a mysterious object.

It is not difficult to understand why this magically elastic, jumping ball, which looked like a stone, would become the object of such wonder and veneration. After all, thanks to improvements in production processes and the development of synthetic materials, it is only very recently that we are able to have smooth, water-resistant balls that can be inflated to the right pressure to enable consistent bounce, balance, shape, trajectory and velocity.

THE INTERNET

Today the phones in our pockets can connect us with friends on the other side of the world, stream a high-definition film, play a brand-new videogame, order our groceries and find information on almost every topic imaginable. All of this is made possible by the internet, an innovation that is now so enmeshed in our everyday lives that we can struggle to imagine life without it. Now the cornerstone of contemporary culture, the technology of the internet has come a long way from its roots as a file-sharing system for government researchers.

The predecessors of today's internet were far smaller networks of computers, which allowed researchers working for a single institution to share data and more efficiently process computing time. One of the most significant examples was the Advanced Research Projects Agency Network (ARPANET), funded by the United States Department of Defence. ARPANET used a process of data communication called "packet switching", a method still used today, which involved breaking down data into smaller sections, or "packets", and transmitting them separately over a network to be reassembled at the other end. ARPANET was the first network to use packet switching over a wide area, enabling computers at universities hundreds of miles apart across the US to communicate with each other via phone lines.

ARPANET was officially declared operational in 1971, and by 1973 satellite links were being made to academic, military and research institutions in Hawaii, Norway and University College London. Within a decade, hundreds of sites would be connected. Not long after the network was established, users realised it was an ideal system for sending messages, and so email was born. Although no one could have realised it at the time, ARPANET was a demonstration of the flexibility of the internet and a sign of the world to come. Personal messages, gossip and newsletters were all being transferred alongside work and research.

"Not long after the ARPANET network was established, users realised it was an ideal system for sending messages."

As more computers from different organisations and nations were added to the network, it became apparent that protocols were needed to ensure compatibility. A core principle in this regard was the transmission-control

The "Magic Modem", a stand-alone modem for the BBC "B" computer, 1981–1982.

protocol, known as TCP/IP, which establishes a common language for computers. We still use a version of this system today: every device connected to the internet has an IP address, which tells other computers where it is in the network.

ARPANET and other early networks like it allowed researchers to communicate quickly, easily and securely. In the 1970s and early 1980s, cheaper technology led to the proliferation of desktop computers, bringing computing out of specialised workplaces like research and finance, and into offices, homes and schools. It also led to the proliferation of local area networks (LANs), which transformed workplaces and initiated the development of some of the first multiplayer games. Among them was *Multi-User Dungeon*, known as *MUD*, created in 1978 by Roy Trubshaw and Richard Bartle on the University of Essex network, before later being connected to ARPANET and becoming the first truly online game. Players explore a fantasy dungeon through text, typing instructions: "walk north", "get sword", and so on.

Although ARPANET would eventually be shut down, it worked as a spine, linking together enough other networks to create what we now call the internet, albeit on a far smaller scale. Computers would use modulator-demodulators, or modems, like the "magic modem" shown on page 217, to convert their digital information into analogue information to send along a phone cable. However, it was not exactly user-friendly and would be unrecognisable to most internet users today.

"The improvement of infrastructure allows data to be transmitted at near lightspeed."

The most recognisable face of the internet, the World Wide Web, did not emerge until 1989. The web was developed as an information management system by British computer scientist Tim Berners-Lee, who was working at the European research organisation CERN at the time. The principle of the web is that documents are full of "hyperlinks", which connect to other documents. All of these documents are created using a special code and have a unique identifying address: this code is called the HyperText Markup Language (HTML) and the address is the Universal Resource Locator (URL). Combined with a piece of software, a web browser, to read and present these pages to the user, all the basic pieces of the World Wide Web are in place. On 6 August 1991, the code to create more web pages and the basic web browser software was made freely available on the internet, and computer enthusiasts began setting up their own collections of pages on servers around the world: websites.

Use of the internet grew quickly, from 2.6 million users in 1990 to 4.7 billion in 2020. In 1990, more than 2 million of those users were in North America alone. Now, most internet users are based in Asia (more than half, in fact), followed by Europe, and then North America. But this is an ever-changing landscape, and the number of internet users in Africa is predicted to overtake those in North America. Yet, the population of Africa is almost equal to that of North America and Europe combined, showing that global internet access still has a long way to go, even before taking into account the speed and reliability of those connections.

When the internet was running on phone lines – the so-called dial-up system – it was limited to a maximum speed of 56 kilobits per second. The fastest nations in the world now have speeds in excess of 250 megabits per second, nearly 4,500 times faster. These increases are down to a number of factors. Changes in protocols and the development of server technology all have a role to play, but the main factor is the improvement of infrastructure, especially the rollout of fibre-optic cable, which allows data to be transmitted at near lightspeed.

We exist in a world where almost everything is connected to the internet, from our health systems and our banks to our home heating and our doorbells. We are more connected than ever before in who we are and what we do, but that also makes us vulnerable to hackers, scams and cyberwarfare.

Cyberwarfare is not the only risk to our welfare and security. With its vast farms of servers around the world, Google estimates its carbon footprint is equal to the nation of Laos. And global internet usage accounts for some 3.7 per cent of all CO_2 emissions, which is roughly equal to the carbon footprint of global air traffic.

The history of the development and spread of the internet is an astonishing story of technological development and cooperation, but more remarkable is how we have used it and the impact of that use. The future is unknowable, but one guarantee is that it will be shaped by the internet.

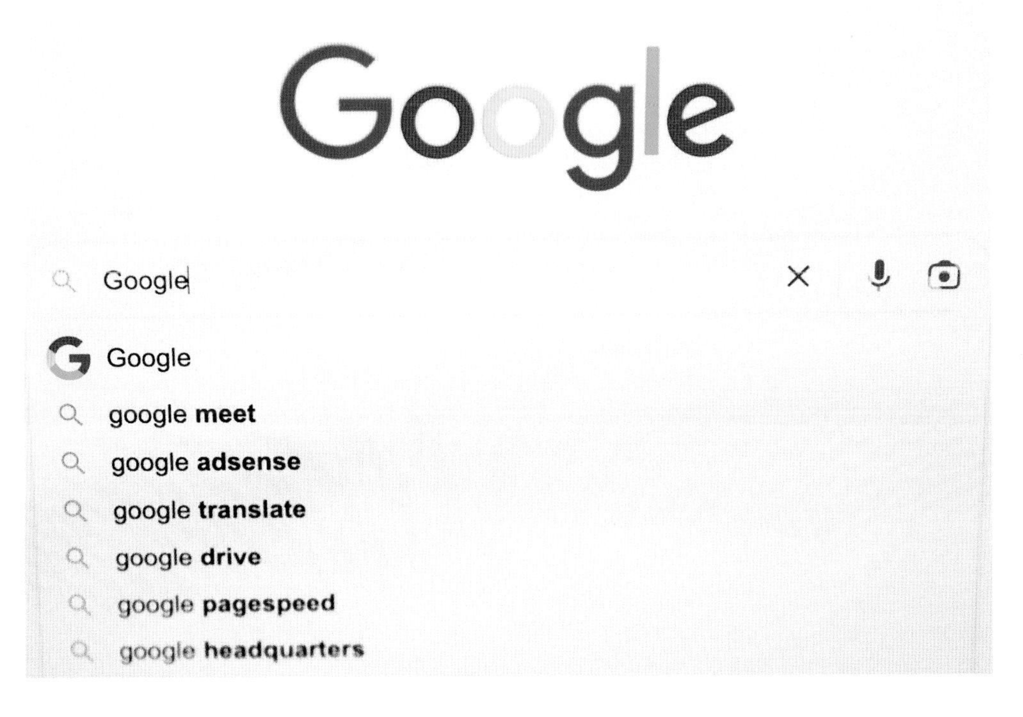

Google is the world's most popular search engine, visited nearly 90 billion times every month.

A pair of dice made in imitation ivory celluloid, early 20th century.

PLASTIC

We live in a plastic world. The artificially created wonder substance has countless uses and has been substituted for other materials, from glass and stone to cotton and animal horn, throughout our homes and daily lives.

The first synthetic plastics were created in the 19th century, and it was British engineer Alexander Parkes who patented the first one in 1855, a synthetic plastic named "Parkesine". Like many of the early plastics, it was derived from a natural material – in this case, plant cellulose. It was treated with nitric acid, then dissolved in alcohol and, finally, hardened into a transparent material, which was mouldable when heated. Including pigmentations into the process allowed the material to resemble ivory, just like the dice shown opposite.

Indeed, a key motivation for the development of synthetic plastics was to find a manufacturable alternative to natural substances such as ivory, animal horn, tusk, tortoiseshell and other similar materials. Adapted and refined by businesses and inventors like American John Wesley Hyatt, who founded the Celluloid Manufacturing Company, these early plastics were able to replace rare materials and make premium objects like billiard balls and combs accessible to a wider audience, and open up mass production options for everyday objects. Celluloid also provided the medium to carry photographic film, revolutionising photography and enabling the birth of cinema.

"A key motivation for the development of synthetic plastics was to find a manufacturable alternative to natural substances."

In the early 20th century, plastic production was revolutionised further with Leo Baekeland's creation of the first entirely synthetic material containing no molecules found in nature. The Belgian chemist had been looking for a synthetic substitute for shellac, an insect resin, to use as an electrical insulator when, in 1907, he combined the chemicals formaldehyde and phenol under heat and pressure to create the new plastic substance, which he named Bakelite. Bakelite's dark, wood-like appearance and easy mouldability lent itself particularly well to mass production, and brought new design trends such as the Art Deco movement to a broad audience, many examples of which remain popular today with lovers of vintage and "retro" household objects.

Production of plastics and the active pursuit of new varieties as substitutes for natural materials continued throughout the century, though this was driven less by a desire to preserve the natural environment and more by a desire to simplify and secure supply chains. This was especially the case during World War II, when the newly invented synthetic silk, nylon, was diverted from its use in toothbrush bristles and women's stockings to be used for parachutes, ropes, body armour and more. Similarly, the glass-substitute Perspex, trademarked as Plexiglas in the US, was used for submarine periscopes and aircraft windscreens.

Not all of the new materials were replacing existing natural alternatives, however. Some seemed entirely new. Teflon, for example, was discovered by accident by chemist Roy J. Plunkett. Its durable, slippery nature made it ideal for coating machine parts and, like nylon and Perspex, it was conscripted during World War II, most notably to coat pipes and valves containing the noxious chemicals needed to enrich uranium for the Manhattan Project. After the war, it found uses in creating waterproof fabrics, easy-clean carpets, fishing line and, most famously, non-stick cookware.

Polyethylene (also known as polythene) was also discovered by accident, when a leak led to oxygen contaminating an experiment to combine the chemicals ethylene and benzaldehyde. In 1935, British scientist Michael Perrin worked out how to replicate the accident in a reproducible way, allowing the mass manufacture of what is now the most common plastic in the world. Polyethylene was first used to insulate radar cabling during World War II, and today more

Our reliance on plastics has had a hugely negative impact on our environment, with plastic products filling landfill sites or finding their way into the oceans.

than 100 million tonnes are produced worldwide annually. It surrounds us, from plastic shopping bags to Tupperware to artificial hips and water pipes.

All of these examples are thermoplastics, materials that become highly pliable and thus mouldable when heated to specific temperatures, but hold their shape and become rigid when they cool. They are also polymers, meaning "many" (*poly-*) "part" (*-mer*), so called because their chemical structure features long repeating chains of molecules. This is why so many plastics have "poly-" names. Behind their brand names, Bakelite, Perspex and Teflon are Polyoxybenzyl-methyleneglycolanhydride, Polymethyl methacrylate, and Polytetrafluoroethylene, respectively. Many of the wonder materials of the 20th century are also a product of the merging of the chemical and petroleum industries, which joined together to search for profitable applications for the waste products of refining crude oil and natural gas. Their successes made companies like ExxonMobil, DuPont and Dow Chemicals household names.

We now know that the ubiquity of plastic is not without its problems. What seemed like benefits of the materials – their durability and the ease and low cost of mass producing them – have turned out to be double-edged swords. In the 20th century, we filled our lives with single-use plastic, using a carrier bag or a plastic bottle once before throwing it away. Discarded to landfill, or as litter in our landscape, these plastic objects can take potentially tens of thousands of years to degrade, and while doing so leach chemicals and small particles (microplastics) into the environment. Plastic bags have been found in the deepest parts of the Mariana Trench, more than 10,000 m (32,810 ft) below sea level, and microplastic particles have been found in the snow of Antarctica, near the peak of Everest and in human blood.

The plastics themselves are not the only problem. In some cases, the chemicals used to create the plastics, or the byproducts of those processes, can be just as worrying. For example, Perfluorooctanoic acid (PFOA) was used in the creation of Teflon products such as non-stick pans. PFOA is an example of what is known as a "forever chemical" with carcinogenic properties, which doesn't break down in the environment or in our bodies. Almost every human on earth is estimated to have some small concentration of PFOA in their bloodstream as a result of its spread into the food chain and water supply. Since 2013, the chemical industry has ceased using PFOA to make kitchenware.

In more recent years, efforts have been made to improve the environmental impact of plastics through better recycling methods and schemes, and more products made from reclaimed plastics. Consumer habits are shifting away from plastic as the go-to solution, looking instead towards other materials, including returning to natural options. Chemists have also worked to develop bioplastics, which biodegrade and break down safely in the environment.

Ultimately, however, plastics are extremely useful materials and remain essential in many parts of our world. And the environmental issues are as much a social problem as they are a scientific one. We need to move away from single-use plastic, and return to treating plastic objects as desirable products we want to use over and over again.

This cinematograph was bought by Save dada, the first Indian filmmaker.

MOVIES

The cinematograph, like the one pictured opposite, was not the first film camera. It is predated by devices such as Thomas Edison's Kinetoscope, and the chronophotographic gun invented by Étienne-Jules Marey. But in terms of its impact, and what it represents in the history of cinema, it is among the most important.

Invented by brothers Auguste and Louis Lumière, French manufacturers of photography equipment, the Cinématographe, or cinematograph, encapsulated the entire cinematic process in its small size, and was first presented at a scientific meeting in March 1895. Unlike Edison's electric device, it was operated with a hand crank, and contained a projector that allowed audiences to view the finished film. By fitting into a box that weighed less than 8 kg (17½ lbs), this compact, portable device freed filmmakers from the confines of the studio, enabling them to make films on location and capture real-world footage, including workers going about their normal day and other spheres of society, while also enabling them to show those films wherever they chose.

For instance, in Paris in December 1895, the Lumière brothers were the first to present projected moving pictures to a paying audience. They screened 10 films, each less than a minute long, many of which would have been unfilmable in a conventional studio, such as workers leaving the Lumière factory, blacksmiths working in a forge, traffic at Lyon's La Place des Cordeliers, and bathers jumping into the sea – all silent and in black and white. Yet, despite laying claim to the first cinema experience, the brothers were initially sceptical about the prospects of their own innovation, allegedly remarking that cinema is "an invention without a future", and declining to sell their creation to eager filmmakers such as Georges Méliès, who attended the screening.

"The Lumière brothers were the first to present projected moving pictures to a paying audience: 10 films, each less than a minute long."

The portability of the cinematograph allowed the brothers to screen their films around the world, in Brussels, London, New York, Buenos Aires, Cairo and Mumbai. The cinematograph pictured here was bought by Indian photographer Harishchandra Sakharam Bhatavdekar, also known as Save dada, who used it to make some of the first Indian films – a demonstration of the rapid global impact and proliferation of the Lumière brothers' technology and the phenomenon of cinema.

By 1914, film was an international industry, dominated by European and Russian filmmakers. As the industry grew, more people paid to see films, and more money was being spent on making them. Professional film studios were established, and the technology of both cameras and projectors became increasingly refined.

World War I greatly disrupted European dominance in the field, however, with many studios and films destroyed alongside the human cost of the war. The studios of French illusionist, actor and film director Georges Méliès, for example, were occupied by the French army and used as a hospital for wounded troops, while hundreds of the studio's films were melted down for raw material. This disruption coincided with the growth of filmmaking in the US, and ultimately led to Hollywood's dominance on a global stage.

Subsequent decades would lead to further innovations in filmmaking technology, not least sound and colour. Sound had always accompanied film, but it was played separately, either by live accompaniment or phonograph records. The first feature-length film to incorporate spoken dialogue – a "talkie", as they were known at the time – was *The Jazz Singer* (1927), which used separate record discs with each reel of film. This was a difficult method to work with, though the film itself was a sensation. Thankfully, a method had already been developed to encode audio information onto the edge of the celluloid: the Movietone sound system. Initially used in newsreel, the system gained popularity and, by the end of 1929, silent movies were the exception rather than the rule in Hollywood.

The sound system also helped make cinemas sites of information. Before widespread television ownership, newsreels at the cinema were a vital source of information. Newsreels both reflected and contributed to more regular film-going in the early decades of the 20th century. They achieved particular importance for British audiences during World War II, providing an important (if heavily monitored and censored) source of news from the front lines and the government. Newsreel filmmakers often undertook acts of astonishing courage to capture footage with unwieldy equipment, entering crumbling, bombed-out buildings or filming near the front lines. Such newsreels nonetheless fell somewhat out of favour as the war progressed as they gained an association of being bearers of bad news.

Like sound, advancements in colour persisted during this period. Early colour films were tinted and coloured, sometimes by hand, an expensive and labour-intensive process. For example, by 1906, the French filmmaker Pathé employed 200 women (dubbed the Pathé hens) to apply colour to their films by hand; it was felt that women were better suited to the fine detail work of creating stencils and applying the coloured dyes. By the early 20th century, the development of specialised techniques and film, such as Kodak's Kodachrome, enabled filming in colour. These methods were complicated and still expensive, but the introduction of a new Technicolor process in 1932 enabled bigger, bolder films like *The Wizard of Oz* (1939) and *Gone with the Wind* (1939) – the latter of which remains the highest-grossing film of all time at the global box office, adjusted for inflation.

The spectacle of sound and colour contributed to the so-called Golden Age of Hollywood in the 1930s and '40s. Ornate "picture palaces" holding thousands of filmgoers per screen were filled night after night, in towns and cities across the United States and around the world, with many people attending the cinema more than once a week. In the UK in 1946, there were 31 million visits to the cinema each week. By 1984, however, that had declined to a million. The growth of out-of-town multiplexes in the 1980s, and the rise of the blockbuster movie, would lead to some growth in numbers in the UK and US once more.

The changing face of cinema has seen lowering numbers in the US and Europe versus their heyday, but growth in other parts of the world: China now boasts the largest cinema attendance in the world, and joins India and Nigeria as the nations now producing more films than the US. *Parasite* (2019), a South Korean film, became the first non-English language film to win Best Picture at the US-dominated Academy Awards. Even as other nations rise in profile, film studios are increasingly transnational companies with editors, visual effects artists, marketing teams, and other sections of the process potentially on the other side of the world from the director, actors and cameras. But for all the ways in which filmmaking has changed over the decades, so too has movie watching.

Today, film screening is overwhelmingly digital, with cinemas playing movies on projectors from hard drives rather than reels of celluloid. And most film watching is now done from the comfort of our homes, a trend that began a succession of physical media from videotape to DVD and Blu-ray. But even this method of viewing has changed and physical media has dwindled in popularity, taking with it specialised businesses previously thought unassailable such as the video-rental company Blockbuster. Films are now predominantly streamed digitally from specialised on-demand services offering menus of content to choose from, old and new, paid for by subscription or one-off rental fees. Facing competition from other media competing for our time, including "bingeable" television shows enabled by those same on-demand services, the film industry is having to change and adapt once again.

The history of cinema is one of change and development, the competing interests of artistic vision, technological change and commercial interest. The future of the medium promises to push these competing factors into still starker relief, with actors and directors worried about the ways streaming business models change the way films are made and paid for, or the unpredictable rise of artificial intelligence and sophisticated effects. At the same time, films are more accessible than ever, global cinema has never been so widely received, and sophisticated filmmaking tools such as high definition cameras and specialist editing software are more affordable than ever. Whatever the future holds for film, one thing remains certain: none of these advancements would have been possible without 19th-century pioneers like the Lumière brothers.

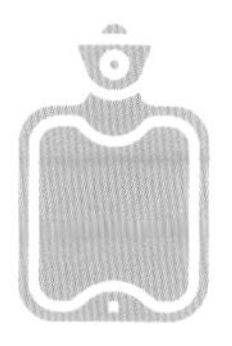

HOT WATER BOTTLES

Encased in colourful, soft covers and held close in times of need – whether for warmth and comfort or to soothe aches and pains – the hot water bottle is a cherished household item for many. It has been one answer to the question of how to keep ourselves warm.

Early attempts to heat up beds involved huge metal pans (generally copper) filled with smouldering coals. Unsurprisingly, the risk of sparks flying was a concern. Metal or stoneware containers with stoppers or caps were introduced in the early 1800s. Their domed shape was designed to spread the heat as much as possible when inserted between bedsheets. Once filled with hot water, they could be incredibly heavy, and any leaks or cracks could lead to burns. Pewter examples came in a range of shapes and sizes, and sometimes with elaborate ornamental engraving, indicating they may have been given as gifts. Metal hot water bottles are still popular in Japan, where they are known as *yutanpo*, and versions are likely to have been in use for centuries.

Eduard Penkala, a Croatian engineer, is often acknowledged as the inventor of the rubber hot water bottle. Known as the "Termofor", a Polish translation of the phrase *hot water bottle*, his invention was patented between 1902 and 1903. However, mentions of hot water bottles appear in the *British Medical Journal* in 1875, and in *The Hospital* in 1893, a London-based journal on health care and health care institutions. One advert reads: "Few are insensible to the comfort a hot water bottle confers, and none are wise who venture between the sheets, when travelling in winter, without the safeguard of such a companion."

"Eduard Penkala is often acknowledged as the inventor of the rubber hot water bottle, patented between 1902 and 1903."

In the US in 1900, Bailey's Good Samaritan hot water bottle was advertised in the following terms: "soft as a pillow", "fits the body" and "soothes and relives" pain. It is likely that these early examples used American chemist Charles Goodyear's development of vulcanised rubber.

Rubber hot water bottle by Haffenden Moulding Co. Ltd, England, 1997.

Note the holes in the copper pan, indicating that this warmer held smouldering coals.

Natural rubber is a milky sap from a variety of trees indigenous to South America and South Asia. When mixed with acid, it is sticky, soft and mouldable. Ingenious peoples across the world have used the material to make containers, balls and water carriers. In North America in the 1830s, Goodyear was trying to strengthen and remove the stickiness of rubber while also making it resistant to extreme temperatures. Several of his early rubber inventions, including a life jacket and mailbags, failed in hot weather. After years of experiments, and learning from Thomas Hancock's methods of heating rubber to strengthen it, Goodyear finally made a breakthrough – by accident. When he dropped his sulphur-rubber mix into a heater, the result was a tough, durable material.

Sourcing rubber to make a whole range of products, including boots, tyres, condoms and hot water bottles, meant searching for the trees in South Asia and South America. Huge competition for rubber led to the creation of plantations, altering landscapes forever. Often under European colonial rule, people and their natural resources were exploited over long periods of time, as newly planted rubber trees take several years to grow.

Demand for rubber remains incredibly high. Nowadays, 85 per cent of the world's natural rubber supply is produced by 6 million smallholders working plots of land in Thailand, Indonesia, China and West Africa. Forests and rainforests have been cleared to make way for new rubber cultivation, putting biodiversity at risk. Rubber trees also need a large water supply. Anything can threaten a rubber tree crop – disease, changing climates and fluctuating prices. Work is ongoing to make rubber production more sustainable and less exploitative.

Finding alternatives to natural rubber involves sourcing different plants or creating synthetic versions. Developing synthetic rubber relied on the chemical understanding of the material, and it took time to replicate. From the 1890s to the 1930s, efforts in Germany developed some of the world's first synthetic rubber. When natural rubber trade routes to the Allies were cut off during World War II, synthetic rubber production was hastened in the US. Much of this was produced for the Allied war effort, and civilians were encouraged to collect as many rubber goods as they could, including hot water bottles. New synthetic materials introduced in the 1930s and '40s, including PVC (polyvinyl chloride), are now a popular choice, particularly for people allergic to latex from rubber.

With more shapes available, a rectangular hot water bottle with a capacity of around 2 litres is the most common. It is possible that rectangular shapes are easier to manufacture, with less waste. Shapes are cut out from rubber of an exact thickness. In a mould, rubber pieces are pressed together under pressure and heat, vulcanised, and cooled quickly. A stopper is added and excess rubber trimmed. Finally, each bottle is air-tested for leaks. Some designs have ribbing to maintain the temperature evenly by creating pockets of air between the bottle and the cover. A smooth side offers more targeted heat.

This classic household item has risen in popularity recently. With rising energy prices in Europe, hot water bottle sales have increased as a potential way to cut costs by "heating the human and not the home". Some UK retailers estimated hot water bottle sales increased by 200 per cent in 2022. Yet while providing comfort, hot water bottles are not without their risks. The British Burn Association estimates that on average one person a day in the UK will experience severe scalding while using or filling a hot water bottle. Some accidents happen while people are asleep, meaning that there is a delay in receiving help. In worst-case scenarios, people experience nerve damage, or second or third degree burns, and may require surgeries and skin grafts.

Hot water bottle covers are one way to create a barrier between your body and the heat source, while also enabling you to individualise this often plain-looking item. Patterns for knitted, crocheted and quilted designs are freely available, and shaped covers (teddy bears, sloths and even unicorns among them) mean there is a selection for those that can afford them.

Another way to prevent harm from hot water bottles is to replace them regularly. Introduced by the British Standards Institute in 2006, the "date daisy" printed on the side tells you when your hot water bottle was made. Each petal represents a month of the year, with dots inside the petal indicating which week of the month it was made. The year of manufacture is at the centre. Knowledge of the date daisy or "daisy wheel" went viral on social media in 2022 and 2023, and many rushed to check theirs having never noticed it before. The recommended time frame for replacement is around three years, but they should be replaced sooner if changes to the rubber or PVC become apparent. Repurposing hot water bottles in creative ways can prolong the material's useful life; they can be cut up into bath toys, filled with wadding and repurposed as cushions, or even turned into jewellery and bags. And rubber and PVC is recyclable at specialist sites.

For as long as there have been rubber or PVC hot water bottles, there have been innovative alternatives. Electric bed warmers were first proposed in the 1920s, often advertised as electric hot water bottles or designed to look like their rectangular rubber counterpart. Connected directly to the mains or a bedside lamp, the electricity heats up an element or heating tube inside. A slightly more unusual example from the early 1900s is Martin's waterless bed-warming bag. When two teaspoons of cold water were added to canvas bags containing an ammonia compound, a chemical reaction produced heat. However, the accompanying smell cannot have made for a pleasant, safe or relaxing experience.

Covering hundreds of years of heating history, hot water bottles have been one popular way of keeping ourselves warm.

CONTRIBUTORS

DK and the Science Museum would like to thank the following colleagues for their contributions and expertise.

William Sims – *Electric taxi* (p.8)
Thimble (p.66)

Rebecca J. I. Mellor – *Condoms* (p.12)
Hair tongs (p.26)

Harriet Jackson – *Baby's bottle* (p.16)
Wheelchair (p.53)

Elissavet Ntoulia – *Cosmetics* (p.21)
Chair (p.75)

Laura Büllesbach – *Matches* (p.30)
Nylon stockings (p.46)

Rebecca Raven – *Period products* (p.36)

Esme Mahoney-Phillips – *Roller skates* (p.40)

Kerry Grist – *Flushing toilet* (p.58)
Lightbulb (p.100)
iPod (p.186)

Katherine McNab – *Selfies* (p.62)
TV sets (p.116)

Angela M – *Mobile phone* (p.70)

Rachel Boon – *Personal computer* (p.80)

Holly Palmer – *Invisible hearing aid* (p.84)
Teasmade (p.112)
Fridge magnet (p.130)

Rebekah Chitson – *Microwave oven* (p.90)

Laura Joy Pieters – *Tinned food* (p.95)

Sarah Bond – *Plasters* (p.104)

Abi Wilson – *Rover safety bicycle* (p.108)

Christopher Valkoinen – *Rulers* (p.120)

Gabrielle Bryan-Quamina – *Contraceptive pill* (p.125)

Rachael Simões – *Toothbrush* (p.134)

Nathan Bossoh – *Kola nut* (p.138)

Heather Bennett – *GPS* (p.143)

Lewis Pollard – Play School *toys* (p.146)
AXBT microphone (p.154)

Charlie Southerton – *Ballpoint pen* (p.150)

Sabrina Ruffino Giummara – *Toy theatre* (p.160)

Katie Crowson – *Locks* (p.166)

Georgina Kavanagh – *Scissors* (p.170)

Sarah Bond – *Pregnancy test* (p.174)

Selina Hurley – *First aid kit* (p.178)
Hot water bottle (p.228)

Matthew Howles – *Teacup* (p.183)

Rachael Mascarenhas – *Sticky tape* (p.192)

Lily Hayward – *Swiss army knife* (p.196)
Kenwood A700 (p.208)

Andrea Lathrop Ligueros – *QWERTY keyboard* (p.200)

Alexandra Rose – *Bath toys* (p.204)

Scott Anthony – *Peruvian rubber ball* (p.212)

Glyn Morgan – *The internet* (p.216)
Plastic (p.220)
Movies (p.224)

INDEX

Index entries in **bold** relate to chapter topics. Page numbers in italics relate to content that is found only in photographs on the page indicated.

ABOUT THE SCIENCE MUSEUM

The Science Museum is part of the Science Museum Group, the world's leading group of science museums that share a world-class collection providing an enduring record of scientific, technological and medical achievements from across the globe. Over the last century the Science Museum, the home of human ingenuity, has grown in scale and scope, inspiring visitors with exhibitions covering topics as diverse as robots, code-breaking, cosmonauts and superbugs. www.sciencemuseum.org.uk.

PICTURE CREDITS

The publisher would like to thank the following for their kind permission to reproduce their photographs:

(Key: a-above; b-below/bottom; c-centre; f-far; l-left; r-right; t-top)

4 Alamy Stock Photo: Oleksiy Maksymenko Photography (tr). **6 Dreamstime.com:** Vectomart (tr). **10 Getty Images:** Heritage Images / Hulton Archive / National Motor Museum. **11 Getty Images:** Handout / Dynamo Taxi (3) / Matt Fowler. **14 Getty Images / iStock:** structuresxx. **15 © Science Museum Group:** Wellcome Institute for the History of Medicine / DUREX (bl, bc, br). **19 Getty Images:** Bloomberg / Jason Alden. **20 Getty Images:** AFP / Frederic J. Brown / Staff. **23 Getty Images:** Science & Society Picture Library. **24 Getty Images:** AFP / Sia Kambou / Stringer. **28 Getty Images:** De Agostini / DEA / W. BUSS. **29 Getty Images:** Hulton Archive / Topical Press Agency / Stringer. **33 Getty Images:** Science & Society Picture Library. **34 Alamy Stock Photo:** GRANGER- Historical Picture Archive. **38 Getty Images:** The Washington Post / Joshua Yospyn. **39 Alamy Stock Photo:** Gioia Forster / dpa. **42 Getty Images:** Russell Lee / Farm Security Administration / Corbis Historical. **43 Getty Images:** The Washington Post / Tom Kelley. **45 Getty Images:** AFP / Mark Ralston / Staff (t); Archive Photos / Interim Archives / Library of Congress (b). **48 Alamy Stock Photo:** Penta Springs Limited. **49 Getty Images:** Hulton Archive / Historica Graphica Collection / Heritage Images. **50 Getty Images:** Hulton Archive / Keystone / Stringer. **51 Getty Images:** SSPL / The National Archives. **52 Threads. 55 Getty Images:** De Agostini / G. Dagli Orti. **57 Getty Images:** Allsport / Adam Pretty / Staff. **64 Getty Images:** Handout / Twitter / Ellen DeGeneres. **69 Alamy Stock Photo:** Science History Images. **72 Alamy Stock Photo:** RBM Vintage Images. **73 Getty Images:** Zhang Peng / LightRocket. **77 Wellcome Collection:** M0003080: Charles Darwin's chair, Down House. **79 Getty Images:** AFP / Yoshikazu Tsuno / Staff. **82 Getty Images:** Corbis Historical. **85 Alamy Stock Photo:** imageBROKER.com GmbH & Co. KG / More information. **86 Alamy Stock Photo:** Hera Vintage Ads. **88 Alamy Stock Photo:** Patti McConville. **89 Alamy Stock Photo:** Nina Hilitukha. **91 Getty Images:** Science & Society Picture Library. **93 Getty Images / iStock:** E+ / stocknroll (tr). **99 Alamy Stock Photo:** 2020 Images (tr). **Getty Images:** Jeffrey Greenberg / Education Images / Universal Images Group (bl). **102 Alamy Stock Photo:** Chronicle (tl); David Cole (tr). **106 Getty Images:** Picture Post / Stringer / Hulton Archive. **107 Getty Images:** Bloomberg / David Paul Morris (tl). **110 Getty Images:** Hulton Archive / Stringer. **114 Alamy Stock Photo:** Chronicle. **123 Shutterstock.com:** Media_Photos. **133 Dreamstime.com:** Andrey Emelyanenko. **136 Getty Images:** De Agostini / DEA / G. Dagli Orti. **137 Getty Images:** Patrick T. Fallon / Bloomberg. **140 Getty Images:** Sygma / John van Hasselt- Corbis. **141 Getty Images:** Royal Geographical Society (with IBG). **142 Getty Images:** Alexander Sayganov / SOPA Images / LightRocket. **148 Getty Images:** Evening Standard / Hulton Archive. **152 Getty Images:** Hulton Archive / Apic. **153 Getty Images:** AFP / Eric Piermont / Staff. **156 Getty Images:** Corbis Historical / adoc-photos. **158 Getty Images:** Visionhaus. **159 Getty Images:** PB Archive / Peter Bischoff / Stringer. **160 Alamy Stock Photo:** INTERFOTO. **163 Alamy Stock Photo:** Lebrecht Music & Arts. **172 Getty Images:** De Agostini / DEA / G. Dagli Orti. **173 Shutterstock.com:** Zamrznuti tonovi. **176 Getty Images:** Lindsey Nicholson / UCG / Universal Images Group. **177 Alamy Stock Photo:** eye35. **180 Getty Images:** Moment / Rosmarie Wirz. **182 Dreamstime.com:** Chernetskaya. **188 Getty Images:** Corbis Historical / George Rinhart. **189 Getty Images:** De Agostini Picture Library. **190 Getty Images:** CBS Photo Archive. **191 Getty Images:** John Keeble. **199 Bridgeman Images. 210 Getty Images:** Hulton Archive / Fox Photos / Stringer. **214 Getty Images:** Moment / Paul Biris. **219 Getty Images:** NurPhoto / Beata Zawrzel. **222 Getty Images:** Milos Bicanski / Stringer. **230 Getty Images:** Sepia Times / Universal Images Group

Cover images: *Front:* **Alamy Stock Photo:** Gareth Dewar cl; **Dorling Kindersley:** Llandrindod Wells National Cycle / Gerard Brown bl; **Dreamstime.com:** Maxim Sergeenkov crb; **Getty Images:** Youst cra; **Shutterstock.com:** LumenSt bc, Photology1971 cr; *Back:* **Alamy Stock Photo:** i creative bl; **Dreamstime.com:** Chernetskaya tr, Radioval bc, Vectomart ca; **Shutterstock.com:** Stock Up cr